HSE

Safety in the installation and use of gas systems and appliances

Gas Safety (Installation and Use) Regulations 1998 as amended

Approved Code of Practice and guidance

AF854142

London: TSO

HSE

tso
a Williams Lea company

Published by TSO (The Stationery Office), part of Williams Lea,
and available from:

Online
www.tsoshop.co.uk

Mail, Telephone & E-mail
TSO
PO Box 29, Norwich, NR3 1GN
Telephone orders/General enquiries: 0333 202 5070
E-mail: customer.services@tso.co.uk
Textphone: 0333 202 5077

Published with the permission of the Health and Safety Executive on behalf of the Controller of His Majesty's Stationery Office.

First published 1994
Second edition 1998
Third edition 2011
Fourth edition 2013
Fifth edition 2018

ISBN 978 0 7176 6674 4

Approved Code of Practice

This Code has been approved by the Health and Safety Executive, with the consent of the Secretary of State. It gives practical advice on how to comply with the law. If you follow the advice you will be doing enough to comply with the law in respect of those specific matters on which the Code gives advice. You may use alternative methods to those set out in the Code in order to comply with the law.

However, the Code has a special legal status. If you are prosecuted for breach of health and safety law, and it is proved that you did not follow the relevant provisions of the Code, you will need to show that you have complied with the law in some other way or a Court will find you at fault.

Guidance

This guidance is issued by the Health and Safety Executive. Following the guidance is not compulsory, unless specifically stated, and you are free to take other action. But if you do follow the guidance you will normally be doing enough to comply with the law. Health and safety inspectors seek to secure compliance with the law and may refer to this guidance.

Printed in the United Kingdom for The Stationery Office.
SD000338 9/25

Contents

Introduction

What this book is about

1 This Approved Code of Practice (ACOP) and guidance gives practical advice to those with responsibilities under:

(a) the Gas Safety (Installation and Use) Regulations 1998 (SI 1998 No 2451) (GSIUR) as amended; and
(b) sections 2(1), 2(2)(c), 3(1) and 3(2) of the Health and Safety at Work etc Act 1974 (the HSW Act) with regards to standards of training in safe gas installation (see Part B, regulation 3 of this document).

2 It has been prepared by the Health and Safety Executive (HSE) in consultation with representatives from the gas industry.

Audience

3 The regulations and the HSW Act place responsibilities on a wide range of people, including those installing, servicing, maintaining or repairing gas appliances and other gas fittings as well as suppliers and users of gas, including certain landlords. They cover a wide variety of premises' gas systems and appliances.

4 This ACOP and guidance is directed at anyone who may have a duty under GSIUR or sections 2(1), 2(2)(c), 3(1) and 3(2) of the HSW Act (with regards to standards of training in safe gas installation, see Part B, regulation 3).

Content and scope of this ACOP

5 GSIUR deals with the safe installation, maintenance and use of gas systems, including gas fittings, appliances and flues, mainly in domestic and commercial premises, eg offices, shops, public buildings and similar places. The regulations generally apply to any 'gas' as defined in the Gas Act 1986 (amended by the Gas Act 1995), apart from any gas comprising wholly or mainly of hydrogen when used in non-domestic premises. The requirements therefore include both natural gas and liquefied petroleum gas (LPG).

6 The Gas Safety (Installation and Use) (Amendment) Regulations 2018 introduce changes to the following regulations (and associated guidance); there are no changes to any ACOP material:

(a) regulation 2(4) – disapplication of compressed natural gas (CNG) filling stations fed by a dedicated supply from the majority of the requirements of GSIUR, bringing them in line with other industrial premises;
(b) regulation 26(9) – inclusion of an existing exemption allowing alternative checks if it is not possible to examine the operating pressure or heat input;
(c) regulation 36(3) and 36A – flexibility in the timing of landlords' annual gas safety checks and determination of the date when the next safety check is due; and

(d) regulation 40A – requirement to carry out a review of the new regulatory provisions contained in regulations 2(4)(g), 26(9)(c) and (ca), 36(3) and 36A.

7 This document also provides (in Part B, regulation 3) ACOP and guidance on standards of training in safe gas installation. This information was previously provided in COP20 *Standards of training in safe gas installation* and **applies to all premises**, including those otherwise excluded from GSIUR.

About ACOPs

8 Approved Codes of Practice are approved by the Health and Safety Executive (HSE) Board with the consent of the Secretary of State (see Appendix 1: Notice of Approval for details).

9 The ACOP describes preferred or recommended methods that can be used (or standards to be met) to comply with GSIUR and the duties imposed by the HSW Act. The guidance provides advice on achieving compliance, or it may give general information, including explanation of the requirements of the law, more specific technical information or references to further information.

10 The legal status of ACOP and guidance text is explained on the copyright page.

Presentation

11 The ACOP text is set out in **bold** and the accompanying guidance in normal type, the text of the Regulations is in *italics*. Coloured borders also indicate each section clearly.

12 Each regulation is preceded by a short factual summary of the main duties imposed by that regulation to help the reader navigate the document.

Interface with other safety legislation

13 The regulations have an interface with requirements under other legislation, as referred to in this ACOP/guidance. Brief summaries of these provisions are given at www.hse.gov.uk/gas/domestic/.

14 Requirements under GSIUR operate within the overall framework of safety controls imposed by the HSW Act and the Management of Health and Safety at Work Regulations 1999 (the Management Regulations) on employers and self-employed people generally.

Terms used in the regulations, ACOP and guidance

15 Words and expressions which are defined in the HSW Act and the Management Regulations have the same meaning in this Code of Practice and guidance unless the context otherwise requires.

16 Certain words and terms are defined in the regulations themselves and details of these can be found in Part A, regulation 2 of this ACOP.

17 The definitions of some terms in GSIUR (eg 'distribution main', 'emergency control', 'gas fittings', 'installation pipework', 'service pipe' and 'service valve') are also used in the Pipelines Safety Regulations 1996 and the Gas Safety (Management) Regulations 1996. Certain terms (eg 'service pipe', 'gas consumer' etc) are also used in the Gas Act 1986 (as amended), where meanings may differ slightly from those in GSIUR. However, this has no effect on how the provisions of

the respective legislation, such as those concerning maintenance of pipes/pipework, are applied.

18 There are additional words and terms that are used throughout this ACOP (such as 'gas engineer' and 'appropriate standard') which, although not defined in the regulations, do have a particular meaning for the purpose of this ACOP. Details of these words/terms can be found in the Glossary.

Application of GSIUR to the apparently self-employed

19 When considering the application of these regulations and ACOP to people working under another's direction, the following should be considered.

20 If people working under the control and direction of others are treated as self-employed for tax and national insurance purposes, they are nevertheless treated as their employees for health and safety purposes. It may therefore be necessary to take appropriate action to protect them. If any doubt exists about who is responsible for the health and safety of a worker this should be clarified and included in the terms of the contract.

21 However, a legal duty under the HSW Act cannot be passed on by means of a contract and there will still be duties towards others under section 3 of the HSW Act. If such workers are employed on the basis that they are responsible for their own health and safety, legal advice should be sought before doing so.

Carbon monoxide (CO) alarms

22 CO alarms are not covered by GSIUR or this ACOP. Although CO alarms are a useful back-up precaution they must not be regarded as a substitute for proper installation and maintenance of gas equipment by a Gas Safe registered engineer. Such alarms should comply with BS EN 50291 (see Further Information) and carry the appropriate conformity marking. CO alarms should be installed, checked, maintained and serviced in accordance with manufacturers' instructions.

PART A General

Regulation 1 Citation and commencement

Regulation 1

(1) These Regulations may be cited as the Gas Safety (Installation and Use) Regulations 1998 and shall come into force on 31st October 1998.

Regulation 2 General interpretation and application

Summary of regulation 2

This regulation defines important terms used in GSIUR. These definitions, together with regulation 2(2)–(7) (application/interpretation), largely determine how requirements are applied in particular circumstances.

Regulation 2(1)

(1) In these Regulations, unless the context otherwise requires –

"appropriate fitting" means a fitting which –

(a) has been designed for the purpose of effecting a gas tight seal in a pipe or other gasway;
(b) achieves that purpose when fitted; and
(c) is secure, so far as is reasonably practicable, against unauthorised opening or removal;

Guidance 2(1)

Appropriate fitting

23 Use of an 'appropriate fitting' is required in several regulations (eg 6(2), 6(3), 16(3), 22(2), 22(3), 26(5), 33(2) and 33(3)) for sealing off a gasway (ie a passage through which gas may pass) to prevent escape of gas, such as from open-ended pipework following disconnection of an appliance, or to seal off the gas supply to an appliance pending completion of work and safe commissioning.

24 Any such fitting should be designed for the purpose (ie not improvised from whatever is at hand, such as a fitting intended to seal other pipework, eg water pipes), and should comply with appropriate standards (see the *Legislative, normative and informative document list*).[1] For instance, suitable screw-in or soldered fittings should be used which securely blank or cap-off open ends of pipes or incomplete gasways in appliances; conventional isolation valves or consumer emergency controls, which might readily be opened by an unauthorised person, must not be used for this purpose.

Regulation 2(1)

"distribution main" means any main through which a transporter is for the time being distributing gas and which is not being used only for the purpose of conveying gas in bulk;

"emergency control" means a valve for shutting off the supply of gas in an emergency, being a valve intended for use by a consumer of gas;

Guidance 2(1)

Emergency control

25 The emergency control is a valve intended, and readily accessible, for use by the consumer (ie end user) of gas. For example, a valve located in a meter room which is locked (for security), and accessible only to a landlord, gas supplier, gas transporter and/or emergency services, cannot be regarded as an 'emergency control'. Where a meter is fitted, the meter control valve may be used as the emergency control, subject to certain conditions – see regulation 9(1).

26 Although there may be more than one emergency control serving a particular premises, it is the outlet of the first emergency control downstream of the distribution main which marks the interface between a 'service pipe' and 'installation pipework' (see definition of 'service pipe' in paragraphs 45–46). There is a similar interface in other related legislation, eg the Pipelines Safety Regulations 1996 (see 'Terms used in the regulations, ACOP and guidance' in the Introduction).

Regulation 2(1)

"flue" means a passage for conveying the products of combustion from a gas appliance to the external air and includes any part of the passage in a gas appliance duct which serves the purpose of a flue;

Guidance 2(1)

Flues

27 Flues (also known as chimneys in standards) come in many forms: they can be natural draught or fan-assisted, open or room-sealed, single or shared, or various combinations of these. For example, they can range from a simple individual natural draught flue to the more complex shared room-sealed natural draught (eg U- or Se-duct). Further information on types of flue design is given in appropriate standards (see the *Legislative, normative and informative document list*).

Regulation 2(1)

"gas" means any substance which is or (if it were in a gaseous state) would be gas within the meaning of the Gas Act 1986 except that it does not include gas consisting wholly or mainly of hydrogen when used in non-domestic premises;

Guidance 2(1)

The definition of 'gas'

28 The definition of 'gas' includes:

(a) methane, ethane, propane, butane, hydrogen and carbon monoxide;
(b) a mixture of two or more of these gases; and
(c) a combustible mixture of one or more of these gases and air.

29 A mixture mainly composed of (a), (b) or (c) and containing another gas (eg a flammable, non-flammable or inert gas not itself covered by the definition) is also included. However, any gas consisting wholly or mainly of hydrogen when used in non-domestic premises (eg a laboratory in the industrial or education sector) is excluded, as is any gas wholly comprised of a substance not listed in 28(a), such as acetylene, oxygen or nitrogen. The substances within the definition are covered irrespective of physical form, eg both liquid and gaseous phases are included (but see paragraphs 30 and 31).

30 GSIUR therefore covers natural gas, liquefied petroleum gas (LPG), methane from coal mines, landfill gas etc, when these products are stored, supplied or used in situations within the scope of GSIUR, eg see definition of 'gas fittings' and 'gas appliance', and regulation 2(4)–(7). Both gas conveyed to premises through a distribution main (normally natural gas) and gas supplied from a storage vessel (mainly LPG) are covered by GSIUR – see regulation 2(3).

Guidance 2(1)

31 Although the definition of gas includes certain liquefied gases, most situations where LPG is likely to be used in liquid form, eg for automotive use or for grain drying, are not covered by GSIUR – see regulation 2(4)–(6).

Regulation 2(1)

"gas appliance" means an appliance designed for use by a consumer of gas for heating, lighting, cooking or other purposes for which gas can be used but it does not include a portable or mobile appliance supplied with gas from a cylinder, or the cylinder, pipes and other fittings used for supplying gas to that appliance, save that, for the purposes of regulations 3, 35 and 36 of these Regulations, it does include a portable or mobile space heater supplied with gas from a cylinder, and the cylinder, pipes and other fittings used for supplying gas to that heater;

Guidance 2(1)

Gas appliance

32 The definition includes a wide range of appliances fuelled by gas. It is implicit that gas has to be stored, supplied or used (basically as a fuel gas) to be covered by GSIUR. The venting to atmosphere of waste gas from coal mines, landfill sites etc is therefore not covered. However, where recovered vent gas from these places is used in appliances at premises subject to the GSIUR, it is covered.

33 The supply of gas to the propulsion system of any vehicle does not come within the scope of the regulations (see exception in regulation 2(6)). Therefore, LPG or natural gas-powered vehicles, including forklift trucks etc, do not need to conform to GSIUR.

Mobile/portable appliances

34 Mobile/portable appliances, where gas is supplied from a cylinder, are not generally covered. Appliances using non-refillable gas containers (eg some camping stoves, cigarette lighters, hairdressing curling tongs etc) are, for instance, not covered. However, a mobile or portable space heater (such as an air heater, gas fire or convector) which is supplied with gas from a cylinder is subject to regulations 3, 35 and 36 (provided it is in premises covered by GSIUR).

35 For instance, this means that appliances such as LPG cabinet heaters provided by landlords must be properly maintained, and checked for safety at least once a year by a person who is, or is employed by, a member of a class of persons approved by HSE under regulation 3, eg a Gas Safe registered gas engineer (see regulations 3, 35 and 36).

36 Work (eg servicing or maintenance) on any gas appliance (including a mobile or portable space heater) that is carried out at premises generally excluded from GSIUR (such as factories) is not covered, irrespective of whether the appliance is intended for use in situations within the scope of the regulations. This is subject to the requirements in regulation 3(8) concerning work on gas fittings in certain vehicles, vessels or caravans (see also paragraphs 97–99).

37 However, similar safety controls under separate legislation may apply in circumstances not covered by GSIUR (see paragraphs 79–80).

Regulation 2(1)

"gas fittings" means gas pipework, valves (other than emergency controls), regulators and meters, and fittings, apparatus and appliances designed for use by consumers of gas for heating, lighting, cooking or other purposes for which gas can be used (other than the purpose of an industrial process carried out on industrial premises), but it does not mean –

(a) any part of a service pipe;
(b) any part of a distribution main or other pipe upstream of the service pipe;

Regulation 2(1)

(c) a gas storage vessel; or
(d) a gas cylinder or cartridge designed to be disposed of when empty;

Guidance 2(1)

Gas fittings

38 This definition effectively excludes from GSIUR any gas fitting which is part of an industrial process carried out on industrial premises, eg gas-fired industrial process plant and furnaces. The exclusion extends to industrial premises not covered by the exceptions (eg concerning factories, mines and quarries) in regulation 2(4). This means, for instance, that bitumen boilers and welding torches are not covered by GSIUR even where they are used at construction sites and similar places which may otherwise fall within scope (see also paragraphs 63–64).

39 Although control devices primarily intended for use by the consumer, eg knobs and switches for controlling the heat or temperature functions of the appliance, are covered by the definition, work on such devices is generally excluded from GSIUR (regulation 2(6)(c) refers).

40 The definition of 'gas fittings' excludes any service pipe (eg as in a network supplying natural gas), but covers service pipework, such as used in the supply of gas from an LPG storage vessel.

Regulation 2(1)

"gas storage vessel" means a storage container designed to be filled or re-filled with gas at the place where it is connected for use or a re-fillable cylinder designed to store gas, and includes the vapour valve; but it does not include a cylinder or cartridge designed to be disposed of when empty;

"gas water heater" includes a gas fired central heating boiler;

"installation pipework" means any pipework for conveying gas for a particular consumer and any associated valve or other gas fitting including any pipework used to connect a gas appliance to other installation pipework and any shut off device at the inlet to the appliance, but it does not mean –

(a) a service pipe;
(b) a pipe comprised in a gas appliance;
(c) any valve attached to a storage container or cylinder; or
(d) service pipework;

Guidance 2(1)

Installation pipework

41 Installation pipework generally connects a meter or emergency control valve to a gas appliance. It includes pipework used to connect an appliance to other installation pipework (ie an 'appliance connector') and any shut-off device (eg isolation valve) at the inlet to the appliance (regulation 26(6) refers). However, any pipe comprised in a gas appliance, ie forming an integral part of an appliance, is excluded. See diagrams of typical installations in Appendix 2.

Regulation 2(1)

"meter by pass" means any pipe and other gas fittings used in connection with it through which gas can be conveyed from a service pipe or service pipework to installation pipework without passing through the meter;

"primary meter" means the meter nearest and downstream of a service pipe or service pipework for ascertaining the quantity of gas supplied through that pipe or pipework by a supplier;

"re-fillable cylinder" means a cylinder which is filled other than at the place where it is connected for use;

Regulation 2(1)

"the responsible person", in relation to any premises, means the occupier of the premises or, where there is no occupier or the occupier is away, the owner of the premises or any person with authority for the time being to take appropriate action in relation to any gas fitting therein;

"room-sealed appliance" means an appliance whose combustion system is sealed from the room in which the appliance is located and which obtains air for combustion from a ventilated uninhabited space within the premises or directly from the open air outside the premises and which vents the products of combustion directly to open air outside the premises;

Guidance 2(1)

Room-sealed appliance

42 The definition of 'room-sealed appliance' is of key importance in the context of regulation 30 which imposes restrictions on types of appliances which may be installed in certain accommodation. In this definition, the term 'premises' refers to the building or structure in which the appliance is located. The definition includes appliances where combustion air is drawn from an uninhabited space, but this option would be precluded for separate safety reasons (and would contravene GSIUR) in certain situations, eg in boats where air may not be drawn from petrol engine spaces.

43 The term 'room-sealed appliance' covers a wide range of appliances, both natural and fanned draught, including those connected to shared flues, eg U- and Se-duct, and 'vertex' systems – see appropriate standards (in the *Legislative, normative and informative document list*).

44 The definition requires combustion products to be vented directly to open air outside premises. Openings should therefore be avoided in any flue serving such an appliance (ie throughout its length inside any building), apart from any opening which is integral and essential to the correct operation of the flue, purpose-designed and properly located for its application (eg a 'vertex' flue in a loft space). It may be necessary to fit a guard around any such flue opening to prevent possible ignition of nearby combustible materials. See also paragraph 26 and Appendix 3, and for appropriate standards, ACOPs, guidance and other relevant information sources see the *Legislative, normative and informative document list*.

Regulation 2(1)

"service pipe" means a pipe for distributing gas to premises from a distribution main, being any pipe between the distribution main and the outlet of the first emergency control downstream from the distribution main;

Guidance 2(1)

Service pipes

45 Although some of the regulations apply to service pipes, detailed requirements for these pipes are covered by the Pipelines Safety Regulations 1996 (see the summary of legislation at www.hse.gov.uk/gas/domestic/). Certain duties, eg concerning maintenance of service pipes, are also imposed under the Gas Act 1986 (as amended by the Gas Act 1995).

46 Service pipes are normally the property of the gas transporter, who needs to be notified before any work is performed on such pipes, including connecting or disconnecting a meter to a service pipe. At least 48 hours' notice needs to be given to the gas transporter and, in certain circumstances, the meter owner (if different) also needs to be notified, eg see schedule 2B of the Gas Act 1986 (which was inserted by the Gas Act 1995).

Regulation 2(1)

"service pipework" means a pipe for supplying gas to premises from a gas storage vessel, being any pipe between the gas storage vessel and the outlet of the emergency control;

"service valve" means a valve (other than an emergency control) for controlling a supply of gas, being a valve –

(a) *incorporated in a service pipe; and*
(b) *intended for use by a transporter of gas; and*
(c) *not situated inside a building;*

"supplier" in relation to gas means –

(a) *a person who supplies gas to any premises through a primary meter; or*
(b) *a person who provides a supply of gas to a consumer by means of the filling or re-filling of a storage container designed to be filled or re-filled with gas at the place where it is connected for use whether or not such container is or remains the property of the supplier; or*
(c) *a person who provides gas in re-fillable cylinders for use by a consumer whether or not such cylinders are filled or re-filled directly by that person and whether or not such cylinders are or remain the property of that person, but a retailer shall not be deemed to be a supplier when he sells a brand of gas other than his own;*

Guidance 2(1)

Gas suppliers

47 Gas suppliers are subject to certain duties under GSIUR (eg to respond to a gas escape under regulation 37). Other regulations refer to the 'supplier of gas' in connection with duties placed on other people (such as for notification of defective or dangerous appliances under regulations 26 and 34, respectively). The meaning of 'supplier' (and allocation of related responsibilities) depends on specific circumstances, as determined by the definition in the regulations – see paragraphs 48–49.

48 In the case of gas (primarily natural gas) supplied to premises through pipes, the 'supplier' is the person who supplies gas through the primary meter (who will also hold the licence required under section 7 of the Gas Act 1986) and bills the consumer.

49 In the case of gas supplied by filling a storage tank (eg LPG) or in cylinders, the meaning of 'supplier' depends on the circumstances:

(a) **Gas supplied direct to a consumer**: Whether the supply is to a storage tank or in cylinders, the 'supplier' is the 'gas company' whose gas is used to fill the tank or cylinders concerned. For the latter, this is normally the company specified on the gas cylinders (see also 49(b)(ii)).
(b) **Gas supplied to an intermediate person**, eg landlord, who provides a gas supply on to 'tenant' consumers: Allocation of 'supplier' duties is as follows:
   (i) **Gas supplied by filling storage vessel**: Where the gas is to be provided by a landlord for the use of tenant(s) in a building or part of a building (eg flat), the LPG company is responsible (regulation 2(2)(b) provides an exception for the landlord in these circumstances). However, where the gas is provided for use by consumers in premises other than buildings (such as caravans), the landlord (eg caravan park operator) is the 'supplier'.
   (ii) **Gas supplied in cylinders**: The gas company attracts supplier-related duties in all cases, but where the landlord provides the gas for use in premises other than buildings (eg caravans, mobile

Guidance 2(1)

homes, residential park homes etc), duties are shared between the gas company and the landlord.

50 In all of these situations, allocation of 'supplier' duties is not affected by who owns the storage tank or cylinders. Also, those involved in intermediate activities only, eg road tanker operators, or retailers/contract fillers of LPG cylinders (who only handle gas of another company's brand), are not regarded as 'suppliers'.

51 Wherever duties are shared by different people, close co-operation and clear definition of agreed responsibilities are essential, to ensure requirements are effectively met and there are no gaps in safety cover, eg concerning response to a gas escape. Although the meaning of 'supplier' provides the basis for allocating related duties under GSIUR, this does not prevent dutyholders making contractual arrangements. Regulation 37(6) specifically provides for a gas supplier to appoint another person/organisation to act as an 'emergency service provider'.

Regulation 2(1)

"transporter" in relation to gas means a person who conveys gas through a distribution main;

"work" in relation to a gas fitting includes any of the following activities carried out by any person, whether an employee or not, that is to say –

- *(a) installing or re-connecting the fitting;*
- *(b) maintaining, servicing, permanently adjusting, disconnecting, repairing, altering or renewing the fitting or purging it of air or gas;*
- *(c) where the fitting is not readily movable, changing its position; and*
- *(d) removing the fitting;*

but the expression does not include the connection or disconnection of a bayonet fitting or other self-sealing connector.

Guidance 2(1)

Work in relation to a gas fitting

52 For the purposes of GSIUR, 'work' includes do-it-yourself activities, work undertaken as a favour for friends and relatives, and work for which there is no expectation of reward or gain, eg voluntary activity for charities. This means that anyone carrying out such work must have the necessary competence, as required by regulation 3(1). However, membership of an HSE-approved class of persons (under regulation 3(3)) is required only by businesses carrying out gas-fitting work.

53 The definition of 'work in relation to a gas fitting' lists specific activities covered by this term, but this list is not exhaustive and other operations may also comprise 'work'. The definition is wide-ranging and includes activities that could affect in any way the gas safety of a gas fitting (whether new or existing), and whether or not it contains gas (eg replacement of combustion seals and appliance controls). In the context of 'work', terms not otherwise defined in GSIUR (or in the HSW Act) bear their normal meaning.

54 'Installation' may refer to either a new or used/second-hand appliance or fitting (see regulation 26(3) and associated guidance) and is not limited to initial installation in particular premises, ie it includes reinstallation (eg following servicing/repair work).

55 'Installation of a gas appliance (or other gas fitting)' will be work for the purpose of these regulations even where the connection is made by a bayonet fitting or other self-sealing connector (the connection of such a fitting being just one part of the installation).

Guidance 2(1)

56 For the purpose of the definition of 'work', 'disconnecting' means physically detaching or uncoupling a fitting (ie which involves breaking into a gasway), rather than simply isolating it by means of a valve or similar device. Both 'connecting' and 'reconnecting' should be understood accordingly.

57 For the purpose of the definition of 'work', readily movable appliances include appliances such as laboratory Bunsen burners and mobile barbecues, which are readily portable. Other appliances, eg free-standing cookers connected by a flexible hose, are not considered to be 'readily movable', but can be moved temporarily, eg to clean the space they normally occupy; this type of activity is not regarded as 'work' within the meaning of GSIUR. (Where an appliance is connected by means of a bayonet fitting, this will need to be disconnected before moving the appliance – such a disconnection is also excluded from the definition of work, as is the connection or reconnection of this type of fitting.)

58 Lighting or relighting an appliance burner/pilot light, eg after temporary disruption or isolation of the gas supply (such as when gas cylinders are changed, or following operation of an over- or under-pressure control device), does not itself comprise work. However, the resumption of a gas supply in these or other circumstances may involve 'work' (eg where purging a gas fitting or appliance of air or gas); but in certain cases checks on appliances under regulation 26(9) are not required (see regulation 26(10)). Any interruption of the gas supply possibly caused by faulty or maladjusted equipment should be thoroughly investigated and remedial action taken, as necessary, in accordance with the requirements for 'work' under the regulations.

59 Regulation 2(6)(c) provides an exception for certain work activities in relation to control devices primarily intended for use by the consumer, eg knobs and switches for controlling the heat or temperature functions of the appliance (see paragraph 74).

Regulation 2(2)

(2) For the purposes of these Regulations –

(a) any reference to installing a gas fitting includes a reference to converting any pipe, fitting, meter, apparatus or appliance to gas use; and

(b) a person to whom gas is supplied and who provides that gas for use in a flat or part of premises let by him shall not in so doing be deemed to be supplying gas.

Guidance 2(2)

Landlords

60 Landlords providing gas for tenants' use in flats or parts of buildings are not 'gas suppliers' for the purpose of GSIUR and do not attract supplier-related responsibilities. However, the position may be different where the gas is for use in any premises which is not a building, eg a caravan. See also definition of 'supplier' in regulation 2(1) and associated guidance. Irrespective of who the gas supplier is, there are separate duties on landlords under regulation 36, eg requiring gas appliances, installation pipework and flues provided for tenants to be maintained in a safe condition and annual safety checks to be carried out (see regulations 36 and 36A).

Regulation 2(3)–(8)

(3) Subject to paragraphs (4) and (5) below, these Regulations shall apply to or in relation to gas fittings used in connection with –

(a) gas which has been conveyed to premises through a distribution main; or

(b) gas conveyed from a gas storage vessel.

Regulation 2(3)–(8)

(4) Save for regulations 37, 38 and 41 and subject to regulation 3(8), these Regulations shall not apply in relation to the supply of gas to, or anything done in respect of a gas fitting at, the following premises, that is to say –

(a) (i) a mine within the meaning of the Mines and Quarries Act 1954 or any place deemed to form part of a mine for the purposes of that Act, or

(ii) a quarry within the meaning of the Quarries Regulations 1999 or any place deemed to form part of a quarry for the purposes of those Regulations

(b) a factory within the meaning of the Factories Act 1961 or any place to which any provisions of the said Act apply by virtue of sections 123 to 126 of that Act;

(c) agricultural premises, being agricultural land, including land being or forming part of a market garden, and any building thereon which is used in connection with agricultural operations;

(d) temporary installations used in connection with any construction work within the meaning assigned to that phrase by regulation 2(1) of the Construction (Design and Management) Regulations 2015);

(e) premises used for the testing of gas fittings;

(f) premises used for the treatment of sewage; or

(g) installations downstream of an isolation valve which –

(i) form a system exclusively used for the compression of gas;

(ii) are primarily used to supply compressed gas to vehicles; and

(iii) incorporate at least one gas compressor which has an electric motor input power rating exceeding 5 kilowatts,[1]

but they shall apply in relation to such premises or part thereof used for domestic or residential purposes or as sleeping accommodation.

(5) Nothing in these Regulations shall apply in relation to the supply of gas to, or anything done in respect of a gas fitting on –

(a) a self-propelled vehicle except when such vehicle is –

(i) hired out in the course of a business; or

(ii) made available to members of the public in the course of a business carried on from that vehicle,

(b) a sea-going ship;

(c) a vessel not requiring a national or international load line certificate except when such a vessel is –

(i) hired out in the course of a business;

(ii) made available to members of the public in the course of a business carried out from that vessel, or

(iii) used primarily for domestic or residential purposes;

(d) a hovercraft; or

(e) a caravan used for touring otherwise than when hired out in the course of a business.

(6) Nothing in these Regulations shall apply in relation to –

(a) the supply of gas to the propulsion system of any vehicle or to any gas fitting forming part of such propulsion system;

(b) the supply of gas to, or anything done in respect of, a bunsen burner used in an educational establishment; or

(c) work in relation to a control device on a gas appliance if –

1 Regulation 2(4) was amended by the Gas Safety (Installation and Use) (Amendment) Regulations 2018.

Regulation 2(3)–(8)

(i) the device is intended primarily for use by a consumer of gas; and
(ii) the work does not involve breaking into a gasway.

(7) These Regulations shall not apply in relation to a gas fitting used for the purpose of training gas fitting operatives in a college or other training establishment, except that paragraphs (1) to (5) and (7) of regulation 3 shall apply to work in relation to a gas fitting carried out by a person providing such training.

(8) These Regulations shall not apply in relation to a gas fitting used for the purpose of assessing the competence of a gas fitting operative at an assessment centre where such assessment is carried out for the purposes of a nationally accredited certification scheme, except that regulation 3(1) and (2) shall apply to work in relation to a gas fitting carried out by a person carrying out such assessment.

Guidance 2(3)–(8)

Premises covered by GSIUR

61 For the purpose of these regulations (except where the context otherwise requires, eg see regulation 2(2)(b)), 'premises' has the meaning given in section 53 of the HSW Act. It covers any place, including any installation on land, the foreshore, offshore installations (whether floating or otherwise – but only within Great Britain), and any tents, marquees or other movable structures.

62 GSIUR therefore covers everywhere in Great Britain, subject to the exceptions for vehicles, ships/boats, factories, agricultural land, construction installations, quarries, mines and other premises/situations in regulation 2(4)–(8). In particular, the regulations cover gas systems and appliances in houses, flats, shops, offices, other commercial and public premises, and holiday accommodation, eg chalets and certain caravans, mobile homes and boats on inland waterways (see paragraphs 63–70).

63 Exceptions to the application of GSIUR are provided for certain premises under regulation 2(4)–(6) as follows:

(a) **Regulation 2(4)**: The places mentioned, eg factories, mines and agricultural premises, are generally excluded, but regulations 37 (gas escapes) and 38 (antifluctuators and valves) do apply and regulation 3(8) specifically covers places where certain caravans and inland waterway boats are manufactured (see paragraphs 96–98).
(b) **Regulation 2(4)(b)**: This excludes certain places, eg docks and electrical power stations, regarded as 'factories' under sections 123–126 of the Factories Act 1961. But gas installation work that is part of a construction activity (also treated as a 'notional' factory under the 1961 Act), eg installation of gas fittings/appliances in buildings being constructed for residential, industrial or commercial use, is covered, with the exception of that in relation to fittings/appliances used in temporary installations such as site huts.
(c) **Regulation 2(4)(g)**: Such installations are either a separate gas supply or an isolatable branch off an existing installation starting at the first isolation valve. This clause covers fittings (including pipework), boosters, compressors, storage and dispensers used primarily for the compression of gas for supply to the fuel tank of a vehicle. However, it does not extend to gas fittings and appliances used for other applications on the same premises (see Figures 1–4).
(d) **Regulation 2(5)–(6)**: Seagoing ships and hovercraft are fully excluded from the scope of GSIUR. Boats, vehicles and caravans may be excluded, depending on the circumstances (see paragraphs 66–72).

Guidance 2(3)–(8)

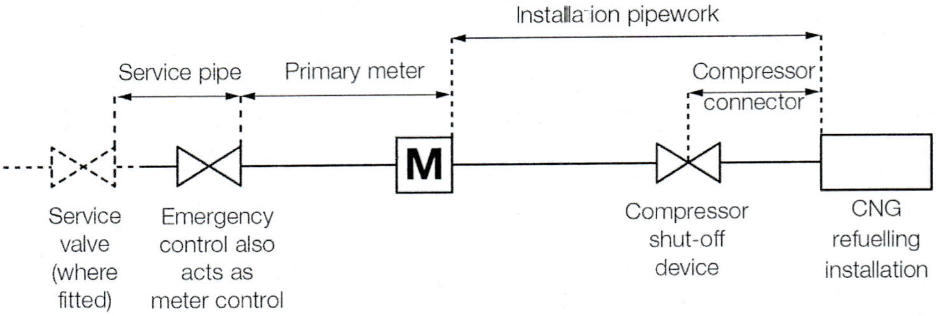

Figure 1 Single metered supply for CNG refuelling: GSIUR doesn't apply

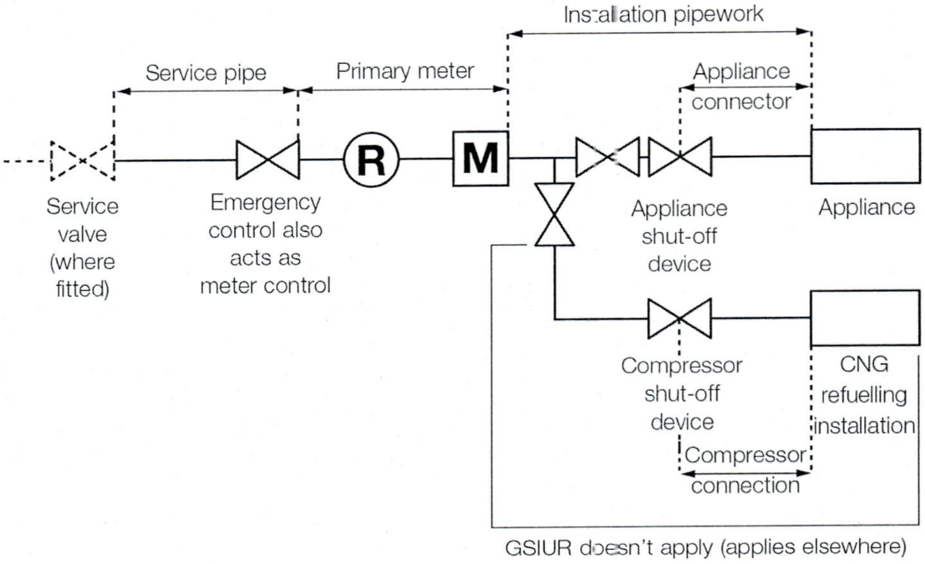

Figure 2 Single metered supply for more than one use

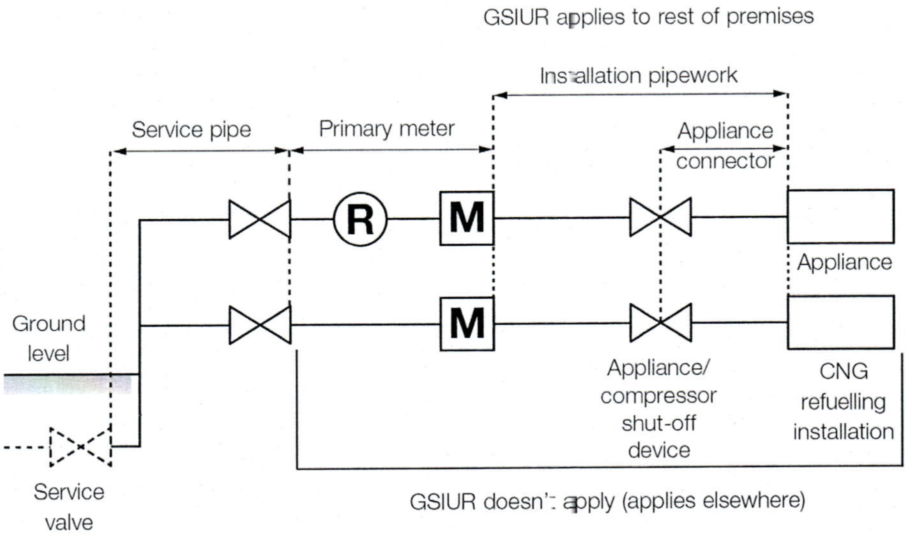

Figure 3 Two separate metered supplies

Guidance 2(3)–(8)

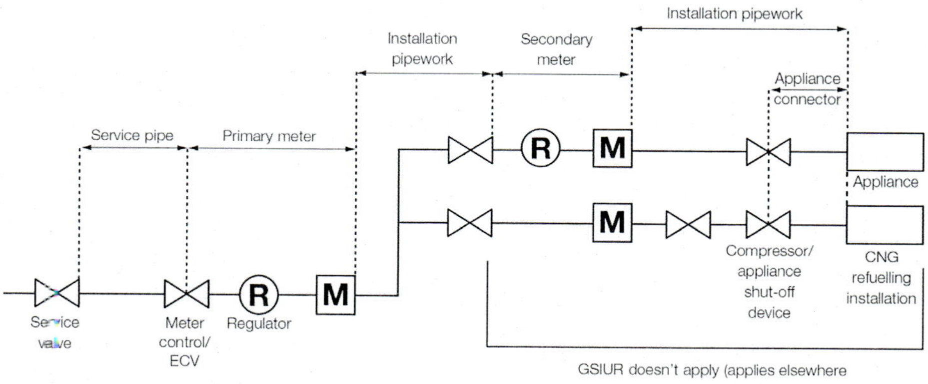

Figure 4 Single supply with two secondary meters

64 The exceptions in regulations 2(4)–(6) cover any gas appliance or fitting (whether fixed or portable) in the premises concerned, irrespective of whether the fitting/appliance is intended for use in a place covered by GSIUR. The range of situations where industrial gas fittings are excluded is further extended (eg to certain other construction situations) by the exclusion for these fittings in the definition of 'gas fittings' (see regulation 2(1), paragraphs 38–40). However, where there are exclusions from these regulations, similar requirements for gas safety may be applied under separate legislation, eg the HSW Act (see paragraphs 79–80).

65 The exceptions in regulation 2(4) do not extend to domestic, residential or sleeping accommodation at the places specified. Therefore, the regulations cover all such accommodation wherever it is found, apart from ships, boats, vehicles etc excluded under regulation 2(5). For example, caretakers' flats at factories or mines are covered.

Caravans

66 Caravans used for touring which are hired out in the course of a business, and permanently sited caravans and mobile homes (hired out or otherwise) on holiday-home parks or similar sites, including farms, are covered. Similarly, any caravan used as an annexe to permanent accommodation, eg as a 'granny flat', or as temporary residential accommodation for employees (for instance on a construction site or farm), is covered by GSIUR.

Boats

67 GSIUR applies to certain boats not requiring a load-line certificate, as specified in regulation 2(5)(c). These include:

(a) boats (both permanently moored and those capable of navigation) which are hired out in the course of a business, eg boats used for holiday accommodation;
(b) any boat (including those privately owned/occupied) used solely or primarily for domestic or residential purposes, such as houseboats and those used for cruising but in which the owner/user lives for most of the time;
(c) boats made available to the public in the course of a business carried out from the vessel, eg floating restaurants and public houses.

68 Privately owned boats used only for leisure or sporting purposes and not hired out in the course of a business are excluded, as are other 'inland waterway

Guidance 2(3)–(8)

boats' not specified in regulation 2(5)(c), eg such boats used as a workplace by an employee of an inland waterway authority for canal repair/clearance purposes etc. Boats requiring a national or international load-line certificate (ie broadly, seagoing cargo-carrying ships) are also excluded from GSIUR, even where used on inland waterways.

69 Vessels subject to GSIUR will often fall outside the scope of merchant shipping legislation (MSL), ie because they are permanently moored and/or not used for commercial purposes. However, in some cases, such as commercially operated pleasure boats (eg hire boats) and certain passenger vessels, GSIUR and MSL apply together. Similarly, additional requirements (including gas safety) may be applied by the navigation authority for the waters on which the vessel operates, for example the Boat Safety Scheme administered by the Environment Agency and the Canal & River Trust, as a condition of issuing licences for use of their waterways.

70 In such situations, compliance with GSIUR will generally satisfy parallel requirements for gas safety under MSL or the relevant navigation authority scheme).[2]

Self-propelled vehicles

71 Self-propelled vehicles hired out to, or frequented by, the public in the course of a business, eg motor caravans, are also covered. However, the tanks and associated fittings of any natural gas or LPG-powered vehicles are excluded under regulation 2(6)(a), eg 'work' on such fittings is not covered (but other safety controls may apply in these circumstances – see paragraphs 79–80).

Mobile food stalls

72 Mobile food stalls (whether towed as 'caravans' or self-propelled) which are made available to the public in the course of a business are covered. This includes such facilities where the public have access inside the vehicle or caravan concerned as well as 'serve-through-the-hatch' vendors. Other types of mobile food stalls (such as marquees, tents, gazebos and portable kitchens) are also covered by GSIUR.

Application in particular circumstances

73 Although GSIUR extends to a wide range of premises, not all uses of gas, or activities in relation to gas fittings/storage vessels, are covered – see, for example, definitions of 'gas appliance', 'gas fittings and 'work' under regulation 2(1). Furthermore, there are specific exceptions from requirements in particular regulations. These factors should all be considered together to determine the application of requirements in particular circumstances.

74 Under regulation 2(6)(c), work on any control device associated with an appliance primarily intended for use by the consumer (eg removal/replacement of electric thermostats and on/off timer switches) is excluded from GSIUR if this does not involve breaking into a gasway, ie where there is no risk of a gas escape. People carrying out such work are not, therefore, required to be members of an HSE-approved class of persons, eg Gas Safe registered (see regulation 3). All the same, competence to ensure any such work is done safely is required under the HSW Act.

75 The exceptions in regulation 2(7)–(8) concerning gas fittings used in the training/assessment of gas-fitting engineers recognise that these fittings may have deliberate faults introduced, eg to demonstrate specific problems or test fault-detection techniques. Most of the regulations (eg concerning appliance safety) do not apply in these circumstances but assessors/trainers are still required to possess the necessary competence to ensure safe conduct of such activities (this

Guidance 2(3)–(8)

needs to take into account the level of trainee knowledge and experience, which might be limited).

76 Training organisations also need to be in membership of an HSE-approved class of persons, eg Gas Safe registered, under regulation 3(3). The exception in regulation 2(7) is restricted to assessment for the purpose of a nationally accredited certification scheme and therefore excludes other 'assessment' activities, eg checks and tests as part of training.

77 Under regulation 2(7)–(8), both assessment and training organisations must ensure that their staff have the necessary competence (see regulation 3(2)). They also need to meet other obligations (eg under the HSW Act, the Management Regulations, and the Workplace (Health, Safety and Welfare) Regulations), for securing safety of both staff and trainees in these circumstances. For instance, this would include addressing activities involving 'faulty' appliances in the risk assessment under the Management Regulations, and implementing the necessary control measures (eg by 'grading' work according to trainee experience, providing appropriate instructions and supervision and implementing formal procedures/checks to ensure gas fittings are left in a 'safe condition' after exercises have been completed).

78 The exceptions in regulation 2(7)–(8) do not cover gas fittings used for purposes other than training/assessment or those used in the training or assessment of engineers at places other than training colleges or assessment centres, eg 'on-the-job' training/assessment carried out at different locations is subject to the regulations in full.

Requirements at premises outside the scope of GSIUR

79 Where there are exclusions under GSIUR (eg at factories) general safety requirements will apply under the HSW Act and related legislation (see summary of legislation at www.hse.gov.uk/gas/domestic/).

80 Even if working at premises to which GSIUR does not apply (eg factories, quarries etc (see regulation 2(4)), competence in carrying out gas work safely is required to comply with the general duties in sections 2 and 3 of the HSW Act. See regulation 3 for details.

PART B Gas fittings – general provisions

Regulation 3 Qualification and supervision

Summary of regulation 3

This section of the ACOP requires anyone carrying out gas work to be competent. Even if working at premises to which GSIUR does not apply (eg factories, quarries etc – see regulation 2(4)), people carrying out gas work must be competent to work safely.

Employers of people carrying out gas work and self-employed people carrying out gas work must be a member of a class of persons approved by HSE.

Regulation 3(1)–(2)

(1) No person shall carry out any work in relation to a gas fitting or gas storage vessel unless he is competent to do so.

(2) The employer of any person carrying out such work for that employer, every other employer and self-employed person who has control to any extent of such work and every employer and self-employed person who has required such work to be carried out at any place of work under his control shall ensure that paragraph (1) above is complied with in relation to such work.

ACOP 3(1)–(2)

81 Gas work should only be undertaken:

(a) by a person who has successfully completed an industry-recognised training course followed by assessment of competence. Training that leads to assessment of competence in safe gas work should be recognised by the industry's standards setting body; or
(b) in the case of a currently or previously registered person, where they have proved competence through a certification scheme; or
(c) for those working at premises that fall outside the scope of the Regulations (see regulation 2(4) and associated guidance), by a person who has successfully completed an appropriate full training course followed by assessment of competence.

82 Training should be of a standard to enable a gas engineer to achieve competence in the safe installation, purging, commissioning, testing, servicing, maintenance, repair, disconnection, modification and dismantling of the gas systems, fittings and appliances with which they are working. This should include an adequate knowledge of:

(a) relevant associated services such as water and electricity;
(b) the potential for exposure to asbestos;
(c) the dangers these may give rise to;
(d) the precautions to take.

Guidance 3(1)–(2)

83 Anyone who does work on a gas fitting or gas storage vessel must be competent to do so (whether or not they are required to be a member of an approved class of persons). Therefore, do-it-yourself gas engineers and those performing favours for friends and relatives all need to have the required competence. The level and range of competence should match the full extent of work done, but needs only to be sufficient for and relevant to that work.

84 Employers of gas engineers are also required under regulation 3(2) to ensure that their employees have the required competence for the work undertaken; in addition to ensuring they are properly experienced and trained (see paragraph 87). This involves ongoing monitoring of performance standards, as necessary.

85 Competence is a combination of practical skill, training, knowledge and experience to carry out the job in hand safely, and ensuring the installation is left in a safe condition for use. Knowledge must be kept up to date with changes in the law, technology and safe working practice.

86 Even if working at premises to which GSIUR does not apply (eg factories, quarries etc – see regulation 2(4)), competence in carrying out gas work safely is required to comply with the general duties in sections 2 and 3 of the HSW Act.

87 The certification schemes referred to in paragraph 81(b) require individual gas engineers to have their competence assessed at regular intervals under specific scheme arrangements set by the standards setting authority, operated by a certification body and currently accredited by the United Kingdom Accreditation Service (UKAS).[3]

88 Further guidance both on the scope of training and the need for proper assessment/reassessment of gas engineers can be found in the industry guidance *Standards of training in gas work*.[4]

89 The duty to ensure gas engineers are competent extends to other employers and self-employed people with control over the work concerned, eg certain contractors and those requiring work to be done in a workplace under their control. Where there is more than one dutyholder in a particular situation, close liaison is essential to ensure requirements are met, eg through proper check procedures.

90 Information on the scope of work a Gas Safe registered engineer is competent to perform may be obtained from the engineer's certificate of competence (issued under the certification scheme referred to in the following paragraphs or from the Gas Safe Register). See also regulation 4 concerning duty to check for membership of an HSE-approved class of persons.

Regulation 3(3)

(3) Without prejudice to the generality of paragraphs (1) and (2) above and subject to paragraph (4) below, no employer shall allow any of his employees to carry out any work in relation to a gas fitting or service pipework and no self-employed person shall carry out any such work, unless the employer or self-employed person, as the case may be, is a member of a class of persons approved for the time being by the Health and Safety Executive for the purposes of this paragraph.

Guidance 3(3)

91 In addition to ensuring competence, all gas engineering businesses, including self-employed gas engineers, are (subject to the limited exceptions in regulation 3(4)) required to be in membership of a class of persons approved by HSE, whether they carry out such work as their main or part activity. Gas engineers who are employed by a member of an approved class of persons but who do separate work on their own behalf need to be in membership of such a class of persons, eg Gas Safe registered, in their own right.

Guidance 3(3)

92 The requirement to be in membership of such a class of persons, eg Gas Safe registered, does not apply to engineers carrying out gas work at premises that fall outside the scope of GSIUR (see regulation 2(4) and associated guidance).

93 At the time of publication of this edition of the ACOP, the only body with such approval is the Gas Safe Register (www.gassaferegister.co.uk), although other organisations may apply to HSE for consideration to act as a registration body if they so wish.

Regulation 3(4)

(4) The requirements of paragraph (3) above shall not apply in respect of –

(a) the replacement of a hose or regulator on a portable or mobile space heater; or

(b) the replacement of a hose connecting a re-fillable cylinder to installation pipework.

Guidance 3(4)

94 Regulation 3 extends to work on portable or mobile space heaters, eg LPG cabinet heaters (where done at premises subject to GSIUR, see paragraphs 34–37).

95 As gas storage vessels are not 'gas fittings' within the meaning of GSIUR, changing cylinders, filling storage tanks or fixing tanks in position does not require membership of an HSE-approved class of persons, eg Gas Safe registration. However, people involved in these activities, or any of those covered by the exceptions in regulation 3(4)(a)–(b) (concerning certain 'like-for-like' hose/regulator replacement activities, such as where used/worn items are replaced) need to have the required competence and ensure that they follow instructions, eg given by the gas supplier/appliance manufacturer, as appropriate.

96 Where other work is involved, eg installing or replacing an LPG tank requiring work on service pipework or other gas fitting(s), it should be carried out only by someone who is in membership of an HSE-approved class of persons, eg a Gas Safe registered engineer, with the required competence. See also regulation 2(6)(c) regarding exceptions for certain work on gas appliance control devices primarily intended for use by the consumer.

Regulation 3(5)–(8)

(5) An approval given pursuant to paragraph (3) above (and any withdrawal of such approval) shall be in writing and notice of it shall be given to such persons and in such manner as the Health and Safety Executive considers appropriate.

(6) The employer of any person carrying out any work in relation to a gas fitting or gas storage vessel in the course of his employment shall ensure that such of the following provisions of these Regulations as impose duties upon that person and are for the time being in force are complied with by that person.

(7) No person shall falsely pretend to be a member of a class of persons required to be approved under paragraph (3) above.

(8) Notwithstanding sub-paragraph (b) of regulation 2(4), when a person is carrying out work in premises referred to in that sub-paragraph in relation to a gas fitting in a vehicle, vessel or caravan –

(a) paragraphs (1), (2) and (6) of this regulation shall be complied with as respects thereto; and

(b) he shall ensure, so far as is reasonably practicable, that the installation of the gas fittings and flues will not contravene the provisions of these Regulations when the gas fittings are connected to a gas supply,

Regulation 3(5)–(8)

except that this paragraph shall not apply where the person has reasonable grounds for believing that the vehicle, vessel or caravan will be first used for a purpose which when so used will exclude it from the application of these Regulations by virtue of sub-paragraphs (a), (c) or (e) of regulation 2(5).

Guidance 3(5)–(8)

97 Regulation 3(8) requires manufacturers of caravans, holiday homes or inland waterway boats for use in situations where GSIUR will apply to ensure employees who install gas fittings (or who do subsequent maintenance or repair work) are competent, and to take reasonable steps to secure the safety of these gas fittings once their products are taken/retaken into use. Manufacturers should take reasonable steps, eg by arrangements with their supply outlets and dealers, to establish, as best they can, the intended use of their products and whether the requirements in regulation 3(8) will therefore apply.

98 Manufacturers also need to bear in mind other more general duties under health and safety and consumer protection legislation. In their own interests, it is recommended that they use the same standards of gas installation in all their products, and have quality control procedures to ensure that this is the case.

99 In addition to the installation of gas fittings themselves, other relevant factors, eg location and size of flue and ventilation openings, need to be taken into account.

Regulation 4 Duty on employer

Summary of regulation 4

This regulation requires any employer or self-employed person requiring work to be done on a gas fitting to take steps to ensure the person doing the work is registered with a class of persons approved by HSE.

Regulation 4

Where an employer or self-employed person requires any work in relation to a gas fitting to be carried out at any place of work under his control or where an employer or self-employed person has control to any extent of work in relation to a gas fitting, he shall take reasonable steps to ensure that the person undertaking that work is, or is employed by, a member of a class of persons approved by the Health and Safety Executive under regulation 3(3) above.

ACOP 4

100 Businesses should make reasonable efforts to obtain evidence that any person they intend to perform gas installation work (subject to regulation 3(3)), either under contract or on their own behalf, is a member, or employed by a member, of a class of persons approved by HSE.

Guidance 4

101 Regulation 4 applies to an employer/self-employed person requiring work to be done on a gas appliance/fitting installed at their workplace as well as to an employer or self-employed person having control over work done on a gas fitting, eg a contractor. The duty supplements that for checking competence under regulation 3(2).

102 At the time of publication of this ACOP, the class of persons approved by HSE are those who are registered with Gas Safe Register (www.gassaferegister.co.uk). Gas Safe Register will provide evidence of registration and confirmation that any certificate of registration is still valid.

Regulation 5 Materials and workmanship

Summary of regulation 5

This regulation imposes a duty on an installer of a gas fitting to ensure that the fitting is suitable for the purpose for which it is to be used. Installation of lead pipe/fittings is prohibited.

Regulation 5(1)

(1) No person shall install a gas fitting unless every part of it is of good construction and sound material, of adequate strength and size to secure safety and of a type appropriate for the gas with which it is to be used.

ACOP 5(1)

103 Gas engineers should acquaint themselves with the appropriate standards for gas fittings and ensure that the fittings they use are to those standards.

Guidance 5(1)

104 Most new gas appliances used for domestic purposes (eg cooking, heating and hot water production) are subject to the Gas Appliances (Safety) Regulations 1995 and will carry the CE marking in addition to any recognised European standard mark and/or BSI kitemark. Some larger plant, eg for commercial use, will also bear the CE marking in conformity with other regulations which implement EC directives, eg on machinery and pressure vessels.

105 Pipes should be of a suitable construction, material, strength and size to convey gas to the appliance and to ensure its safe operation. Consider any special factors in particular circumstances, eg risk of physical damage to non-metallic pipework, for instance, from rodent attack (see also regulations 5(2) and 7). Where the appropriate standard or the manufacturer recommends a limited lifetime for a gas fitting (including plastic pipework), it should be replaced before the end of its limited life, unless it can be shown that continued use will not constitute a hazard. The manufacturer should be consulted to establish whether the lifetime of the gas fitting can be extended, what tests need to be carried out, and the extended lifetime allowed. See also regulations 35–36 and associated guidance concerning maintenance/safety check duties.

Regulation 5(2)

(2) Without prejudice to the generality of paragraph (1) above, no person shall install in a building any pipe or pipe fitting for use in the supply of gas which is –

(a) made of lead or lead alloy; or

(b) made of a non-metallic substance unless it is –

(i) a pipe connected to a readily movable gas appliance designed for use without a flue; or

(ii) a pipe entering the building and that part of it within the building is placed inside a metallic sheath which is so constructed and installed as to prevent, as far as is reasonably practicable, the escape of gas into the building if the pipe should fail.

ACOP 5(2)

106 Non-metallic connectors for use with readily movable gas appliances should conform to the appropriate standard. Such hoses, pipes and pipe fittings must be used only for the purpose for which they are designed.

Guidance 5(2)

107 Appliances or meters may be connected to existing lead piping using suitable fittings, provided that the piping is in a safe condition, eg there is no sign of damage.

Guidance 5(2)

108 Apart from connectors to readily movable appliances such as Bunsen burners, non-metallic (eg plastic) pipes should only be used within buildings if sheathed in metal to minimise the risk of gas escaping if the pipe should fail. This requirement does not extend to polyethylene (PE) piping buried in ground beneath a building provided that the piping does not enter the building (but see regulation 19(5)). However, it does apply to that routed in spaces under floors.

109 Plastic pipe fittings should not be used to seal pipes within buildings (see also definition of 'appropriate fitting' in regulation 2(1)). Although regulation 5(2) applies only to pipes and pipe fittings in buildings, it is recommended that a similar approach be adopted for other premises, eg caravans and inland waterway boats. See also regulation 26(2) concerning pipes for flued domestic appliances.

110 Free-standing gas cookers are generally movable, but are not regarded as 'readily movable' for the purpose of this regulation (see also paragraph 57). Flexible cooker hoses should therefore be of metallic construction; they should conform with appropriate standards and be used only for their intended purpose.

Regulation 5(3)

(3) No person shall carry out any work in relation to a gas fitting or gas storage vessel otherwise than in accordance with appropriate standards and in such a way as to prevent danger to any person.

Guidance 5(3)

111 This regulation covers two separate but related requirements. Work in relation to any gas fitting or gas storage vessel needs both to be in accordance with appropriate standards, and to be carried out in a manner which does not expose the gas engineer, or any other person, to danger, eg from a gas escape.

112 It is the particular work being carried out which needs to comply with the appropriate standards. It is not necessary to upgrade existing installations to bring them up to current standards, eg modifying them to meet revised specifications (unless necessary to meet other provisions of GSIUR).

113 The term 'appropriate standards' is not defined in the regulations themselves, but details of where to find a list of standards regarded as appropriate standards for the purpose of this ACOP and guidance is given in GSIUR (see also relevant paragraphs under 'Terms used in the regulations, ACOP and guidance' in the Introduction).

114 *The Gas Industry Unsafe Situations Procedure*[5] provides details of situations that may present a danger (such as gas leaks and incomplete combustion of gas etc).

Regulation 6 General safety precautions

Summary of regulation 6

This regulation specifies measures to be taken by any person working on a gas fitting against danger from gas release, and requirements for sealing gasways and testing gastightness after work is completed. Requirements are specified for safe installation of gas storage vessels (eg LPG), and the storage of natural gas is prohibited.

Regulation 6(1)–(2)

(1) No person shall carry out any work in relation to a gas fitting in such a manner that gas could be released unless steps are taken to prevent the gas so released constituting a danger to any person.

Regulation 6(1)–(2)

(2) No person carrying out work in relation to a gas fitting shall leave the fitting unattended unless every incomplete gasway has been sealed with the appropriate fitting or the gas fitting is otherwise safe.

ACOP 6(1)–(2)

115 Appropriate fitting(s) used to seal appliance gasways should ensure that the gas supply to the appliance cannot be readily restored until safe to do so. Closure of an emergency control does not fulfil this requirement (see definition of 'appropriate fitting' in regulation 2(1)).

Guidance 6(1)–(2)

116 This regulation generally covers work done on existing gas fittings (including extension or modification of existing gas installations) that involves breaking into a live gasway. If the work has to be left unfinished for any period – during which appliances might be deliberately or inadvertently used – the gas supply to the (disconnected) appliance should be sealed, as specified above. Reference should be made to the requirements for an 'appropriate fitting' under the definition in regulation 2(1).

117 'Unattended' has differing meanings depending on the size of premises. For domestic premises it normally means that the person working on the gas system is absent from the premises for a significant time during which someone else (eg the appliance owner) could turn the gas supply back on to use the appliance. For most larger commercial locations 'unattended' could mean being on another part of the site.

Regulation 6(3)

(3) Any person who disconnects a gas fitting shall, with the appropriate fitting, seal off every outlet of every pipe to which it was connected.

ACOP 6(3)

118 The open end(s) of any metal pipe(s) from which a gas fitting has been disconnected should be sealed with an appropriate metal fitting. See definition of 'appropriate fitting' in regulation 2(1).

Guidance 6(3)

119 This regulation is largely intended to deal with situations in which gas appliances are removed, for instance because they are no longer needed or are being taken away when owners move home. When an appliance is disconnected and open-ended pipework is left, the pipework should always be left sealed with an appropriate fitting.

Regulation 6(4)

(4) No person carrying out work in relation to a gas fitting which involves exposing gasways which contain or have contained flammable gas shall smoke or use any source of ignition in such a manner as may lead to the risk of fire or explosion.

Guidance 6(4)

120 'Sources of ignition' include such tools as blow-lamps and hot-air guns. A safe system of work (including effective isolation of the gas supply etc) should be used to avoid any risk of fire or explosion. Gasways should generally be purged of gas before using tools such as a blow torch.

Regulation 6(5)

(5) No person searching for an escape of gas shall use any source of ignition.

Guidance 6(5)

121 This prohibition extends to householders and other members of the general public.

122 The source of leaks should be located by sense of smell, gas detection instruments approved for use in flammable atmospheres, leak detection fluids, pressure test equipment or a combination of these methods.

Regulation 6(6)

(6) Where a person carries out any work in relation to a gas fitting which might affect the gas tightness of the gas installation he shall immediately thereafter test the installation for gas tightness at least as far as the nearest valves upstream and downstream in the installation.

ACOP 6(6)

123 The gastightness of the installation should be tested in accordance with the appropriate standard.

Guidance 6(6)

124 The relevant standards can be found in the *Legislative, normative and informative document list*.

125 The gastightness test, sometimes referred to as the soundness, or pressure decay test, is used for confirming that the gas system has no gas escapes/leaks. In domestic situations the test normally needs to cover the installation pipework between the meter control valve or emergency control valve as appropriate, and the appliance shut-off device.

Regulation 6(7)

(7) No person shall install a gas storage vessel unless the site where it is to be installed is such as to ensure that the gas storage vessel can be used, filled or refilled without causing a danger to any person.

Guidance 6(7)

126 Guidance on installation of LPG vessels is given in UKLPG Codes of Practice.[6]

127 LPG tanks should not be located indoors, within a bunded area or on a roof of a building. Location of LPG storage tanks should take into account the need for safe access for road tankers where practicable, with provision of off-road parking, and for safe delivery of LPG. Any LPG storage tank used to supply gas to multiple consumers should be located only at a site where the person providing the gas supply to consumers has full control over the access to the tank and associated equipment.

128 In Scotland, the restrictions on the siting of LPG vessels are imposed by the Building Regulations (see summary of legislation at www.hse.gov.uk/gas/domestic/).[7]

Regulation 6(8)–(9)

(8) No person shall install in a cellar or basement:

(a) a gas storage vessel; or
(b) an appliance fuelled by liquefied petroleum gas which has an automatic ignition device or a pilot light.

(9) No person shall intentionally or recklessly interfere with a gas storage vessel or otherwise do anything which might affect a gas storage vessel so that the subsequent use of that vessel might cause a danger to any person.

Guidance 6(8)–(9)

129 The prohibitions extend to everyone, including householders and other members of the public.

130 The prohibitions in regulation 6(8) address the fire/explosion hazard, eg from possible leakage and accumulation of heavy LPG vapour in any cellar or basement or from involvement of a gas storage vessel in a building fire. Further guidance on the design, installation and testing of LPG piping systems and maintenance of LPG installations is given in the UKLPG Codes of Practice.

Regulation 6(10)

(10) No person shall store or keep gas consisting wholly or mainly of methane on domestic premises, and, for the purpose of this paragraph, such gas from time to time present in pipes or in the fuel tank of any vehicle propelled by gas shall be deemed not to be so stored or kept.

Guidance 6(10)

131 This regulation prohibits storage and keeping of natural gas at domestic premises, including for the purpose of home filling of gas-fuelled vehicles. However, this prohibition does not extend to natural gas in pipes, or in fuel tanks of parked or garaged vehicles. Similarly, equipment used for filling vehicle tanks is not prohibited where this does not involve storage, but any requirements stipulated by the gas transporter in relation to antifluctuators/valves need to be met in these circumstances – see regulation 38.

132 The meaning of 'domestic premises' in regulation 6(10) is that given in section 53 of the HSW Act and includes any garden, yard or garage of such premises not used in common by the occupants of more than one such dwelling.

Regulation 7 Protection against damage

Summary of regulation 7

This regulation requires gas fittings to be protected from damage, including physical damage and corrosion, and from blockage by a foreign body, eg dirt/dust.

Regulation 7(1)

(1) Any person installing a gas fitting shall ensure that it is properly supported and so placed or protected as to avoid any undue risk of damage to the fitting.

ACOP 7(1)

133 Gas fittings should be installed in accordance with appropriate standards.

Guidance 7(1)

134 The relevant standards can be found in the *Legislative, normative and informative document list*.

135 Gas fittings should be properly supported to ensure that no undue strain is placed on them (eg because of the weight or action of the gas fitting itself), which could lead to failure or damage of the fitting (see also regulation 31). Gas fittings should, as far as possible, be installed in a position which avoids any risk of damage from foreseeable activities in the vicinity, eg traffic. Where such risk cannot be avoided, fittings should be suitably protected against possible damage. This would need to be addressed, for instance, in the case of service pipework, meter installations and installation pipework, where location near to vehicle access routes cannot be avoided.

136 Details of suitable arrangements for support and protection of gas fittings are given in appropriate standards (see the *Legislative, normative and informative document list*). Appliances need to be installed in accordance with manufacturers' instructions; the Gas Appliances (Safety) Regulations 1995 require such instructions to accompany appliances supplied or placed on the market. See summary of legislation at www.hse.gov.uk/gas/domestic/. Special care needs to be taken where a gas fitting is located in a movable structure such as a caravan or boat, to take account of wind, tides etc.

Regulation 7(2)

(2) No person shall install a gas fitting if he has reason to suspect that foreign matter may block or otherwise interfere with the safe operation of the fitting unless he has fitted to the gas inlet of, and any airway in, the fitting a suitable filter or other suitable protection.

Guidance 7(2)

137 Filters or other suitable protection may be necessary to protect against any adverse effect of dirt or dust. Such devices should be compatible with any connected appliance. If there is any doubt, the advice of the gas supplier/ transporter and/or fitting manufacturer, as appropriate, should be sought.

Regulation 7(3)

(3) No person shall install a gas fitting in a position where it is likely to be exposed to any substance which may corrode gas fittings unless the fitting is constructed of materials which are inherently resistant to being so corroded or it is suitably protected against being so corroded.

Guidance 7(3)

138 Environments of particular concern are those where water, salt-spray and damp, corrosive chemicals or soot are present or likely to be present. Special care needs to be taken where a gas fitting is located in a movable structure such as a boat, or a caravan on coastal sites. In some circumstances it may be necessary to apply a protective coating but that coating should not affect safe operation of the fitting.

Regulation 8 Existing gas fittings

Summary of regulation 8

This regulation prohibits anyone from carrying out any alteration to premises in which a gas fitting or storage vessel is already installed which could compromise gas safety. The regulation also prohibits any work on a gas fitting or associated flue/ventilation system which would result in danger to any person.

Regulation 8(1)

(1) No person shall make any alteration to any premises in which a gas fitting or gas storage vessel is fitted if that alteration would adversely affect the safety of the fitting or vessel in such a manner that, if the fitting or the vessel had been installed after the alteration, there would have been a contravention of, or failure to comply with, these Regulations.

Guidance 8(1)

139 This regulation embraces a wide range of physical alterations to premises that might affect the safety of an existing gas fitting or gas storage vessel installed in the premises where the alteration is to be made (see paragraph 146). Before a significant alteration to premises where a gas appliance is installed (eg installation or removal of windows, air bricks, extractor fan units etc) any implications for gas appliance/fitting safety should be properly addressed (see below).

140 Similarly, prior consideration should be given to the possible effect of any modification, such as a building extension, on the safety of a gas storage vessel (eg from any reduction in separation distances and standards of ventilation), or on service pipework (for example, the risk of damage to buried pipework from strain/ weight imposed on it). See also regulations 6(7)–(8), 7(1) and 19(5).

141 Requirements of Building Regulations, eg concerning ventilation and flues, should also be taken into account. Reference should also be made to regulation 36(11)–(12) concerning duties on landlords in respect of any room to be converted into sleeping accommodation.

142 Regulation 8(1) applies to householders, builders, gas engineers etc, and is supplementary to any requirements under the Construction (Design and

Guidance 8(1)

Management) Regulations 2007. The implications of any change to premises for safety of a gas fitting or gas storage vessel need to be considered systematically as part of the overall work planning process.

143 In some cases, the effects of particular work for gas safety may be obvious. For example, where a chimney is being reduced in height or capped, its effectiveness in removing flue gases may be drastically affected. Before such work is started, it needs to be established whether or not the chimney is active and, if so, allowance made for this in the way the work is carried out. This may include planning for appliances to be disconnected before work commences, preventing debris falling or being thrown down the chimney, and ensuring appliances are tested for safety after the work has been completed. (Disconnection/testing of appliances should only be carried out by someone who is, or is employed by, a member of an HSE-approved class of persons under regulation 3, eg a Gas Safe-registered engineer.) Similar consideration needs to be given to any proposed alterations which might affect operation of a flue system, eg fitting a flue liner or terminal.

144 In other cases, the effects of the work may not be so obvious. For example, when fitting double glazing or cavity wall insulation, the removal of fixed permanent ventilation, such as air bricks, and replacement with closeable ventilation louvres (in contravention of standards) or blockage of vents by insulation material, can lead to danger from reduced ventilation/incomplete combustion. Equally, fitting extractor fans can lead to the pull on flues being overcome and flue products being sucked back into premises. Moreover, the enclosure of an existing flue terminal within a new extension or conservatory (again in contravention of standards) can lead to flue gases becoming trapped.

145 Irrespective of how obvious the implications of the work are for gas safety, the people involved need to keep these matters in mind, and ensure they are properly addressed when alterations to premises are planned.

146 Regulation 8(1) applies only to the alteration of premises in which the gas appliance concerned is installed, ie it does not cover any alteration to adjoining premises. Gas appliances/storage vessels should be installed so that safety cannot be affected by any future developments (eg building extensions on adjoining property. For example, a flue should be located at a safe distance from any site boundary, so that in event of any development on a neighbouring site (which could extend up to the boundary), the flue will continue to operate properly and discharge of combustion products will not present a hazard to anyone. (See regulation 27 and Appendix 3, and for appropriate standards, ACOPs, guidance and other relevant information sources, see the *Legislative, normative and informative document list*.)

Regulation 8(2)

(2) No person shall do anything which would affect a gas fitting or any flue or means of ventilation used in connection with the fitting in such a manner that the subsequent use of the fitting might constitute a danger to any person, except that this paragraph does not apply to an alteration to premises.

Guidance 8(2)

147 This regulation applies to everyone, not just gas engineers. It supplements regulation 8(1) by prohibiting other activities (ie except alteration to premises) which have the potential to compromise safety. This might, for instance, include modifications which cause blockage/obstruction of an air supply vent or flue, or incorrect installation of an air extraction or condensation control unit. As with regulation 8(1), it is essential for the implications of any such change or modification to be properly addressed before work is commenced, to ensure that gas safety cannot be prejudiced in any way.

Guidance 8(2)

148 Modification of any gas fitting should be made only by a competent person who is, or is employed by, a member of an HSE-approved class of persons (eg Gas Safe registered) under regulation 3(3). Alterations not comprising 'work' on a gas fitting but which nevertheless may affect gas safety (eg a change to the provision of ventilation in a room) should also be made only by a competent person. Similarly, any significant modification needs to be checked by a competent person before the gas fitting concerned is taken into use, to ensure that appropriate standards have been met and safety has not been compromised (see regulation 26(9)).

Regulation 8(3)

(3) In relation to any place of work under his control, an employer or a self-employed person shall ensure, so far as is reasonably practicable, that the provisions of paragraphs (1) and (2) are complied with.

Guidance 8(3)

149 It is recommended that a responsible person, eg a principal contractor, foreman, site manager or other person with overall control of the work, is nominated to ensure, so far as reasonably practicable, compliance with regulation 8(1) and (2).

Regulation 9 Emergency controls

Summary of regulation 9

This regulation requires an emergency control to be provided when gas is first supplied to premises. Where this control is not adjacent to a meter, a notice must be posted adjacent to the control describing the procedure in the event of a gas escape.

Regulation 9(1)

(1) No person shall for the first time enable gas to be supplied for use in any premises unless there is provided an appropriately sited emergency control to which there is adequate access.

Guidance 9(1)

150 Whenever a new gas supply is made available for use in premises, an emergency control should also be provided. Where there is a gas meter, the meter control may serve as the emergency control as long as the following conditions are met:

- (a) Each individual premises (eg each house, flat, maisonette, or caravan) using a supply of gas should be provided with an emergency control, whether or not that premises contains a gas meter.
- (b) The emergency control should be situated as near as is reasonably practicable to the point where the gas supply pipe enters the premises.
- (c) It should be readily accessible to all consumers, ie gas users, in the premises concerned (eg not located in a basement or cellar).
- (d) A valve located in a meter room which is normally locked, and accessible only to a landlord, gas supplier, gas transporter and/or emergency services for example, cannot act as an 'emergency control'.
- (e) An emergency control should be protected against unauthorised operation (ie tamper-proof) but if situated in a locked compartment, the occupier(s) of the premises should be provided with keys (see also regulation 13(3)–(4)). In such cases, the emergency service provider should also hold keys where access cannot be ensured for them at all times, eg through keys held by the responsible person for the premises.

151 The person allowing the flow of gas to the premises should ensure that every gas consumer in the premises is aware of the location of their emergency control, and of the action to be taken in case of a gas emergency. (Where there is more

Guidance 9(1)

than one emergency control, eg in multi-occupancy premises, it is important for the particular control serving that consumer to be identified.) This includes following any alterations to buildings or ground works affecting the accessibility of the emergency control and pipework. In the case of rented property, the responsible person for the building, such as a landlord or managing agent, should ensure that all tenants are made aware of this information, including any new tenants (see also definition of 'emergency control' in regulation 2(2)).

152 Some premises contain separate buildings, eg a domestic property with stables or an office complex with a number of buildings on one site. As long as the premises are clearly under the control of one person or organisation, only one emergency control is required. However, arrangements should be made to ensure that the person in control is immediately notified of any gas emergency in the premises, so that suitable action can be taken, eg to isolate the gas supply (see regulation 37).

153 In the case of LPG, this regulation only applies where the gas is supplied from a storage tank or tanks, or from two or more cylinders connected by an automatic changeover device (see regulation 9(5)). In other cases the vapour valve on each cylinder also functions as a shut-off control and no additional provision is necessary.

Regulation 9(2)

(2) Any person installing an emergency control shall ensure that –

(a) any key, lever or hand-wheel of the control is securely attached to the operating spindle of the control;
(b) any such key or lever is attached so that –
  (i) the key or lever is parallel to the axis of the pipe in which the control is installed when the control is in the open position; and
  (ii) where the key or lever is not attached so as to move only horizontally, gas cannot pass beyond the control when the key or lever has been moved as far as possible downwards;
(c) either the means of operating the key or lever is clearly and permanently marked or a notice in permanent form is prominently displayed near such means so as to indicate when the control is open and when the control is shut; and
(d) any hand-wheel indicates the direction of opening or closing of the control.

Guidance 9(2)

154 The emergency control can operate by a key, lever or hand-wheel which should be securely attached to the operating spindle. Where a key or lever is used, the 'open' position should be when the key or lever is parallel to the axis of the pipe. The 'off' position should be approximately one quarter turn of the key or lever to the right or left and, where the key or lever moves in the vertical plane, the move to the 'off' position should be in a downwards direction. Either the key or lever itself, or a nearby permanent notice, should indicate how the control operates and when the gas is 'off' and 'on'.

Regulation 9(3)

(3) Where a person installs an emergency control which is not adjacent to a primary meter, he shall immediately thereafter prominently display on or near the means of operating the control a suitably worded notice in permanent form indicating the procedure to be followed in the event of an escape of gas.

Guidance 9(3)

155 Where an emergency control is installed which is not adjacent to a primary meter, or where no meter is installed, a prominently displayed notice on or near the control bearing the words 'Gas emergency control' should be provided. The notice needs to tell the consumer:

Guidance 9(3)

(a) to shut off the supply of gas if there is a gas escape in the premises;
(b) if gas continues to escape, immediately notify the Gas Emergency Freephone Number 0800 111 999 (if natural gas). If the gas is not natural gas, notify the supplier emergency service. A contact list for LPG suppliers can be found in *The Gas Industry Unsafe Situations Procedure*;
(c) not to reinstate the supply until remedial action has been taken by a competent person to prevent gas escaping again;
(d) details of the emergency gas service contact, including the emergency telephone number. In the case of natural gas, the Gas Emergency Freephone Number should be specified (as above). For any other gas, the name and telephone number of the supplier emergency service should be given (see the contact list in *The Gas Industry Unsafe Situations Procedure*);
(e) the date the notice was first displayed.

156 The notice should be in a permanent form, ie of durable material, and protected against damage as necessary, eg weather-resistant.

157 Where an emergency control is installed adjacent to a primary meter, an emergency notice must be displayed instead at the meter (see regulation 15).

158 See regulation 37 and associated guidance concerning action required in the event of a gas escape.

Regulation 9(4)

(4) Where any person first supplies gas to premises where an emergency control is installed, he shall ensure that the notice required by paragraph (3) above remains suitably worded or shall, where necessary, forthwith amend or replace that notice so as to give effect to the provisions of that paragraph.

Guidance 9(4)

159 In the case of LPG, if there is a change to the gas supplier which results in a change to the emergency contact point, the notice needs to be updated by the new supplier.

Regulation 9(5)

(5) This regulation shall not apply where gas is supplied in a refillable cylinder except where two or more cylinders are connected by means of an automatic change-over device.

Regulation 10 Maintaining electrical continuity

Summary of regulation 10

This regulation requires electrical continuity to be maintained during work on a gas fitting where necessary to avoid danger.

Regulation 10

In any case where it is necessary to prevent danger, no person shall carry out work in relation to a gas fitting without using a suitable bond to maintain electrical continuity until the work is completed and permanent electrical continuity has been restored.

ACOP 10

160 A temporary continuity bond to the appropriate standard is required when disconnecting and reconnecting pipework where the production of a spark could cause a hazard.

Guidance 10

161 The purpose of the temporary bond is to maintain electrical continuity between the pipework to reduce the risk of a spark causing ignition. Work is not considered to have been completed until permanent electrical continuity has been restored where this is necessary to prevent potential hazard. The bond should be fitted with a robust clip or clamp at each end which provides effective electrical contact. See also regulation 18(2) concerning equipotential bonding which addresses the risk of electric shock.

PART C Meters and regulators

Summary of regulations 11 and 12

Regulations 11 and 12 require that gas meters are installed so as to avoid, as far as reasonably practicable, adverse effects on the means of escape from premises, and specify requirements concerning the construction of certain meters. Other requirements are imposed for meter installation, eg to avoid electrical hazards and facilitate inspection/maintenance, and for pipe connections, gastightness tests and purging of meters.

Although regulations 11 and 12 deal with gas safety of meters, there are other safety issues that must be considered, eg electricity or the risk of exposure to asbestos when replacing old meters. Further guidance is available at www.hse.gov.uk.

Regulation 11 Interpretation of Part C

Regulation 11

In this Part –

"meter box" means a receptacle or compartment designed and constructed to contain a meter with its associated fittings;

"meter compound" means an area or room designed and constructed to contain one or more meters with their associated fittings;

"secondary meter" means a meter, other than a primary meter, for ascertaining the quantity of gas provided by a person for use by another person.

Regulation 12 Meters – general provisions

Regulation 12(1)–(2)

(1) No person shall install a meter in any premises unless the site where it is to be installed is such as to ensure so far as is reasonably practicable that the means of escape from those premises in the event of fire is not adversely affected.

(2) No person shall install a meter in any premises unless it is of sound construction adequate to ensure so far as is reasonably practicable that in the event of fire gas is not able to escape in hazardous quantities, save that this paragraph shall not apply to any meter installed in non-domestic premises to which gas is supplied through a readily accessible service valve.

ACOP 12(1)–(2)

162 Before any meter is installed, a check should be made on the means of escape from the premises in the event of fire. Meters and meter compartments, and their installation, must conform to the appropriate standard.

Guidance 12(1)–(2)

General

163 Building designers and contractors need to consult the gas supplier/transporter to agree the position and housing for primary meters at an early stage in any new building development.

164 In the context of paragraphs 165–167, 'new meter' is any installation of a meter in a position where there has not previously been a meter, and 'replacement meter' is the installation of a meter in the position of an existing meter.

Installing new meters (premises with two or more floors above the ground floor)

165 Meters should not be sited on or under the stairway or in any other part of the premises where the stairway or that other part of the premises forms the sole means of escape in the case of fire.

Installing other new meters (including premises with less than two floors above the ground) and replacement meters

166 Meters need to be installed, where reasonably practicable, in accordance with paragraph 165. If it is necessary to install a new or replacement meter on or under a stairway or in any other part of premises where the stairway or that other part of the premises forms the sole means of escape in the case of fire, then:

(a) the meter should be fire-resistant; or
(b) the meter should be housed in a fire-resistant compartment with automatic self-closing doors; or
(c) the pipe immediately upstream of the meter, or regulator if fitted, should be provided with a thermal cut-off device which is designed to automatically cut off the gas supply if the temperature of the device exceeds 95°C.

167 Due account should be taken of relevant requirements in the Building Regulations. It may be necessary to install meters in meter rooms or compounds.

Regulation 12(3)

(3) No person shall install a meter unless the installation is so placed as to ensure that there is no risk of damage to it from electrical apparatus.

ACOP 12(3)

168 Meters should only be installed in a location that ensures that the necessary separation distances between the gas meter installation and electrical apparatus (as specified in the appropriate standards) are maintained.

Regulation 12(4)–(6)

(4) No person shall install a meter except in a readily accessible position for inspection and maintenance.

(5) Where a meter has bosses or side pipes attached to the meter by a soldered joint only, no person shall make rigid pipe connections to the meter.

(6) Where a person installs a meter and the pipes and other gas fittings associated with it, he shall ensure that –

(a) immediately thereafter they are adequately tested to verify that they are gas tight and examined to verify that they have been installed in accordance with these Regulations; and
(b) immediately after such testing and examination, purging is carried out throughout the meter and every other gas fitting through which gas can

Regulation 12(4)–(6)

then flow so as to remove safely all air and gas other than the gas to be supplied.

ACOP 12(4)–(6)

169 Connections should be tested for gastightness in accordance with the appropriate standard. The whole installation should then be purged including any appliances connected to the system, again in accordance with the appropriate standard.

Regulation 13 Meter housings

Summary of regulation 13

This regulation specifies the requirements for meter housings concerning safe dispersal of any gas escape, avoidance of combustible materials, and provision of keys to enable consumer access.

Regulation 13(1)–(2)

(1) Where a meter is housed in a meter box or meter compound attached to or built into the external face of the outside wall of any premises, the meter box or meter compound shall be so constructed and installed that any gas escaping within the box or compound cannot enter the premises or any cavity in the wall but must disperse to the external air.

(2) No person shall knowingly store readily combustible materials in any meter box or meter compound.

Guidance 13(1)–(2)

170 To prevent gas leakage entering the building, any cable or installation pipe should be adequately sleeved and sealed where it passes from the meter box and enters the building (see also regulations 14(1) and 19).

171 Regulation 13(2) applies to everyone, not just gas engineers. Meter boxes or compounds should not, for example, be used for storing combustible household waste such as paper and cardboard.

Regulation 13(3)–(4)

(3) No person shall install a meter in a meter box provided with a lock, unless the consumer has been provided with a suitably labelled key to that lock.

(4) No person shall install a meter within a meter compound which is capable of being secured unless the consumer has been provided with a suitably labelled key for that compound.

Regulation 14 Regulators

Summary of regulation 14

This regulation stipulates the protection arrangements for maintaining gas pressure within safe limits. Requirements are also included for sealing regulators against unauthorised interference.

It includes separate stipulations relating to systems supplied from gas storage tanks (eg LPG) or from certain cylinder configurations.

Regulation 14(1)

(1) No person shall install a primary meter or meter by pass used in connection with a primary meter unless –

(a) there is a regulator controlling the pressure of gas supplied through the meter or the by pass, as the case may be, which provides adequate

Regulation 14(1)

automatic means for preventing the gas fittings connected to the downstream side of the regulator from being subjected to a pressure greater than that for which they were designed;

(b) where the normal pressure of the gas supply is 75 millibars or more at the inlet to the regulator, there are also adequate automatic means for preventing, in case the regulator should fail, those gas fittings from being subjected to such a greater pressure; and

(c) where the regulator contains a relief valve or liquid seal, such valve or seal is connected to a vent pipe of adequate size and so installed that it is capable of venting safely.

Guidance 14(1)

172 An over-pressure shut-off device or pressure relief valve may be used to prevent exposure of downstream fittings to excessive pressure in the event of regulator failure. Requirements for protection against over-pressure in the case of gas installations supplied from vessels or cylinders (eg LPG) are given in regulation 14(2)–(4).

173 Any vent pipe should discharge to open air rather than into any part of a building, and should terminate away from potential ignition sources. Reference should be made to the appropriate standards.

174 No installation pipework or cables should directly enter the property from the meter box where the normal pressure of the gas supply is 75 millibars or more at the inlet to the regulator, and the meter is housed in a meter box or meter compound attached to or built into the external face of the outside wall of any premises (see also regulation 13(1)).

Regulation 14(2)

(2) Without prejudice to the requirements of paragraph (1), no person shall cause gas to be supplied from a gas storage vessel (other than a re-fillable cylinder or a cylinder or cartridge designed to be disposed of when empty), to any service pipework or gas fitting unless –

(a) there is a regulator installed which controls the nominal operating pressure of the gas;

(b) there is adequate automatic means for preventing the installation pipework and gas fittings downstream of the regulator from being subjected to a pressure different from that for which they were designed; and

(c) there is an adequate alternative automatic means for preventing the service pipework from being subjected to a greater pressure than that for which it was designed should the regulator referred to in sub-paragraph (a) above fail.

Guidance 14(2)

175 Regulation 14(2) applies only to installations using gas (essentially LPG) stored in bulk vessels or tanks. Any such installation should be provided with:

(a) a regulator to maintain the gas supply pressure within the range at which appliances downstream were designed to operate safely; and

(b) an over-pressure and under-pressure shut-off device (OPSO/UPSO) to provide back-up protection if the regulator fails or the gas supply pressure falls to a dangerously low level, eg because the storage vessel has become empty.

176 In addition, where there is a risk of service pipework failing if the regulator malfunctions, the pipework should be protected by a further over-pressure shut-off device to prevent such failure. This is normally only required for non-metallic pipework.

177 Installed pressure systems and associated gas storage vessels should comply with the Pressure Systems Safety Regulations 2000, where applicable (see

Guidance 14(2)

summary of legislation at www.hse.gov.uk/gas/domestic/). Any regulator or over- or under-pressure protection device required under regulation 14(1)–(4) should be designed, constructed and installed in accordance with the appropriate standards.

Regulation 14(3)–(4)

(3) No person shall cause gas to be supplied through an installation consisting of one or more re-fillable cylinders unless the supply of gas passes through a regulator which controls the nominal operating pressure of the gas.

(4) Without prejudice to paragraph (3) above, no person shall cause gas to be supplied through an installation consisting of four or more re-fillable cylinders connected to an automatic change-over device unless there is an adequate alternative means for preventing the installation pipework and any gas fitting downstream of the regulator from being subjected to a greater pressure than that for which it was designed should the regulator fail.

Guidance 14(3)–(4)

178 Regulation 14(3)–(4) applies only to installations using gas (essentially LPG) from cylinders. As with installations using bulk vessels, a regulator should be provided to protect appliances downstream from being subjected to a pressure outside their safe operating range. If the cylinder supply comprises four or more cylinders connected to an automatic changeover device, an over-pressure protection device should be fitted to prevent downstream appliances and pipework being subjected to a pressure above that for which they were designed, should the regulator fail. Automatic cylinder changeover devices should comply with the appropriate standard.

179 Regulation 14(4) outlines the minimum legal requirements. However, it is recommended that all fixed pipework cylinder installations have an over-pressure protection device fitted to prevent downstream appliances, pipework and fittings from being subjected to a pressure above that for which they are designed.

Regulation 14(5)–(7)

(5) Where a person installs a regulator for controlling the pressure of gas through a primary meter, a meter by pass used in connection with a primary meter or from a gas storage vessel, or installs a gas appliance itself fitted with a regulator for controlling the pressure of gas to that appliance, he shall immediately thereafter ensure, in either case, that the regulator is adequately sealed so as to prevent its setting from being interfered with without breaking of the seal.

(6) In relation to –

- *(a) gas from a distribution main, no person except the transporter or a person authorised to act on his behalf;*
- *(b) gas from a gas storage vessel, no person except the supplier or a person authorised to act on his behalf,*

shall break a seal applied under paragraph (5) above other than a seal applied to a regulator for controlling the pressure of gas to the appliance to which that regulator is fitted.

(7) A person who breaks a seal applied under paragraph (5) shall apply as soon as is practicable a new seal which is adequate to prevent the setting of the regulator from being interfered with without breaking such seal.

Guidance 14(5)–(7)

180 Regulation 14(6) generally restricts the breaking of regulator seals to the gas transporter or supplier in the particular situation, or someone authorised to act on their behalf. However, a competent person may adjust a regulator on an individual appliance, eg to check the correct operating pressure, without the need for authorisation to break the seal.

Guidance 14(5)–(7)

181 Any person who breaks a seal should ensure that it is replaced with a suitable equivalent seal as soon as the work has been completed.

Regulation 15 Meters – emergency notices

Summary of regulation 15

This regulation requires that an emergency notice is posted at a primary meter giving the procedure to be adopted in the event of a gas escape. A notice showing the position of the emergency control is also required in certain cases.

Regulation 15(1)–(2)

(1) No person shall supply gas through a primary meter installed after the coming into force of these Regulations or for the first time supply gas through an existing primary meter after the coming into force of these Regulations unless he ensures that a suitably worded notice in permanent form is prominently displayed on or near the meter indicating the procedure to be followed in the event of an escape of gas.

(2) Where a meter is installed or relocated in any premises in either case at a distance of more than 2 metres from, or out of sight of, the nearest upstream emergency control in the premises, no person shall supply or provide gas for the first time through that meter unless he ensures that a suitably worded notice in permanent form is prominently displayed on or near the meter indicating the position of that control.

ACOP 15(1)–(2)

182 Any supplier of gas through a primary meter should ensure that the required emergency notice is in place. Where there is a change in gas supplier which involves a change in the emergency service provider, the notice should be updated/replaced to reflect such a change.

Guidance 15(1)–(2)

183 Whenever a primary meter is installed in premises, an emergency notice should be prominently displayed on or near the meter. The notice should contain the information for the gas consumer, as specified in paragraph 155.

184 If the nearest upstream emergency control is installed in or subsequently relocated to a position out of sight of the meter or in a position more than 2 metres away from the meter, a notice should be displayed on or near the meter concerned indicating the position of the emergency control.

185 Notices required under regulation 15 should be in a permanent form, ie of durable material, and protected against damage as necessary, eg weather-resistant. The notice(s) should not obscure other information shown on a meter. See regulation 37 and associated guidance concerning action required in the event of a gas escape.

186 Where an emergency control is not located adjacent to a primary meter, an emergency notice is required instead to be displayed on or near to the emergency control (see regulation 9(3)).

Regulation 16 Primary meters

Summary of regulation 16

This regulation prohibits the installation of a prepayment meter as a primary meter in certain cases. It also specifies requirements for notices at primary meters where gas is supplied to more than one secondary meter. Precautions, eg for isolation/sealing, are specified for situations where a primary meter has been removed.

Regulation 16(1)–(2)

(1) No person shall install a prepayment meter as a primary meter through which gas passes to a secondary meter.

(2) Any person –

(a) who first provides gas through any service pipe or service pipework after the coming into force of these Regulations to more than one primary meter; or

(b) who subsequently makes any modification which affects the number of primary meters so provided,

shall ensure that a notice in permanent form is prominently displayed on or near each primary meter indicating that more than one primary meter is provided with gas through that service pipe or service pipework.

ACOP 16(1)–(2)

187 This regulation prohibits the installation of both traditional prepayment meters and smart meters switched to prepayment mode as primary meters where gas is passed on to a secondary meter(s).

Guidance 16(1)–(2)

188 Regulation 16(1) prohibits the installation of a prepayment meter (either a traditional prepayment meter or a smart meter used in prepayment mode) as a primary meter, where gas is passed on to a secondary meter(s), because use of such a meter would (when money runs out) cause the gas supply to all consumers on the system to be immediately cut off. This would lead to a potential risk of gas escape where any gas appliance has not been isolated before the gas supply is resumed, eg because the user is away at the time (see also paragraph 195).

189 Although regulation 16(1) itself only applies to the installation of a prepayment meter, the general duties will apply, and so following installation, no person should change the mode of use of a primary meter (where gas is passed to secondary meter(s)) to prepayment mode. This applies to both traditional meters and smart meters.

190 The purpose of the notice required under regulation 16(2) is to make gas engineers, emergency services etc aware of the possible implications for other users, eg if the gas supply needs to be isolated. The notice is required both for new installations and where an existing system is modified so that more than one primary meter is then supplied from a service pipe or service pipework. In the case of modifications, the person making the change is responsible for ensuring the notice is displayed. The notice should also indicate the number and location of other meters supplied from the same service pipe or pipework, and that any person who carries out an alteration affecting the accuracy of the information provided should ensure that the notice is amended accordingly.

Regulation 16(3)–(4)

(3) Where a primary meter is removed, the person who last supplied gas through the meter before removal shall –

(a) where the meter is not forthwith re-installed or replaced by another meter –

(i) close any service valve which controlled the supply of gas to that meter and did not control the supply of gas to any other primary meter; and

(ii) seal the outlet of the emergency control with an appropriate fitting; and

(iii) clearly mark any live gas pipe in the premises in which the meter was installed to the effect that the pipe contains gas; and

(b) where the meter has not been re-installed or replaced by another meter before the expiry of the period of 12 months beginning with the date of

Regulation 16(3)–(4)

removal of the meter and there is no such service valve as is mentioned in sub-paragraph (a)(i) above, ensure that the service pipe or service pipework for those premises is disconnected as near as is reasonably practicable to the main or storage vessel and that any part of the pipe or pipework which is not removed is sealed at both ends with the appropriate fitting.

(4) Where a person proposes to remove a primary meter he shall give sufficient notice of it to the person supplying gas through the meter to enable him to comply with paragraph (3).

Guidance 16(3)–(4)

191 Regulation 16(3)(b) interfaces with the duty on pipeline operators under the Pipelines Safety Regulations 1996 (PSR) to ensure that certain disused pipelines are left in a safe condition. In the case of any gas service pipe (covered by PSR), the gas supplier (as defined in regulation 2(1)) should ensure, so far as reasonably practicable, that the action stipulated in regulation 16(3)(b) is taken by the dutyholder under PSR (ie the gas transporter).

192 In the case of pipework not covered by PSR, eg an LPG system on a caravan or mobile home site, the 'gas supplier' should take the necessary action to disconnect/seal pipework or make appropriate arrangements for this to be done. In both cases the work should be carried out only by a person meeting the requirements of regulation 3.

193 Regulation 16(3)(b) requires disconnection 'as near as is reasonably practicable' to the main or storage tank. However, the regulation recognises that there are certain situations, eg blocks of flats, where removal of service pipes or service pipework to particular dwellings may not be readily achievable. The gas supplier (together with the transporter in the case of natural gas) needs to decide the appropriate action in a particular case, bearing in mind that unused 'live' service pipes or pipework represent a potential hazard. Metal fittings should be used to seal metal pipes (and emergency control outlets under regulation 16(3)(a)). Any installation pipework or other gas fitting rendered redundant by meter removal should be made safe (eg by purging, and sealing any open gasways with an appropriate fitting).

194 Regulation 16(4) recognises that meters may be removed by persons other than gas suppliers. Close co-operation with the 'last supplier' of gas is essential in these circumstances. Sufficient notice of any proposed meter removal should be given to enable the supplier to take effective action under regulation 16(3), as required by schedule 2B of the Gas Act 1986, as amended.

Regulation 17 Secondary meters

Summary of regulation 17

This regulation requires that any person supplying or permitting the supply of gas through a primary meter to a secondary meter (eg a landlord) must display, at specified positions, a notice showing the configuration of the gas system.

Regulation 17(1)–(2)

(1) Any person supplying or permitting the supply of gas through a primary meter to a secondary meter shall ensure that a line diagram in permanent form is prominently displayed on or near the primary meter or gas storage vessel and on or near all emergency controls connected to the primary meter showing the configuration of all meters, installation pipework and emergency controls.

Regulation 17(1)–(2)

(2) Any person who changes the configuration of any meter, installation pipework or emergency control so that the accuracy of the line diagram referred to in paragraph (1) is affected shall ensure that the line diagram is amended so as to show the altered configuration.

Guidance 17(1)–(2)

195 This regulation applies where a gas supply is provided through a primary meter to secondary meter(s), ie to a number of downstream consumers. In the case of multiple-occupation houses or blocks of flats, an engineer or the emergency services would require information on the location/configuration of emergency control valves and pipework to isolate the gas supply to a particular dwelling, and to establish how this would affect the supply to other consumers in the premises. Similarly, this information would be required when supplies to individual consumers needed to be checked, eg after disconnection or reconnection of pipework.

196 The situation may be further complicated in 'sub-deduct' situations, ie where secondary meters are used by one or more gas supplier(s) for establishing charges to downstream consumers by a process of deduction from the total quantity of gas supplied through a primary meter (see the Gas Act 1995 – Network Code, section G1). A configuration may be complex, involving several sub-deduct meters, associated pipework and emergency controls, together with the primary meter. Clear and unambiguous information on these configurations is essential, for instance, to avoid situations where action to isolate the supply to one consumer unknowingly interrupts someone else's supply (leading to possible risk of a gas escape where appliances are not isolated before the supply is reinstated, eg because the user is away at the time).

197 The notice must contain the information specified in regulation 15(1). A diagram showing the configuration in outline will normally be sufficient for this purpose, ie a detailed scale plan is not necessary. The diagram should be displayed in a prominent position at the primary meter (or gas storage vessel if the gas is so supplied) and at all emergency controls connected to the primary meter in the premises. The duty to provide the notice(s) rests with the person supplying or permitting the supply of gas through the primary meter. This might be a gas supplier (eg in sub-deduct situations) or a landlord providing gas to a tenant. Where there is more than one dutyholder, effective liaison is essential to ensure requirements are met.

198 Regulation 17(2) requires any person making a change to the configuration of any installation pipework, meter or emergency control to ensure the notice originally provided under regulation 17(1) is suitably amended, where this is necessary to preserve accuracy for safety purposes, as described above. The notice does not need to be modified in respect of minor changes which do not have implications for safety identification purposes, eg where there is a small change in location of a relevant fitting but relative positions are unaffected.

199 The advice of a competent gas engineer should be sought on the wording of the notice. In addition to the information specified in regulation 17, it is recommended that notices indicate the person responsible for each section of installation in the premises. The notice may be based on the system diagram required under regulation 24, where applicable.

PART D Installation pipework

Regulation 18 Safe use of pipes

Summary of regulation 18

This regulation requires pipework to be installed in a safe position, having regard to factors which might affect safety, eg location of other pipes, drains, cables and electrical apparatus. Any person connecting installation pipework to a meter is required to inform the person responsible for the premises (eg the occupier) of the need for equipotential bonding.

Regulation 18(1)–(2)

(1) No person shall install any installation pipework in any position in which it cannot be used with safety having regard to the position of other pipes, pipe supports, drains, sewers, cables, conduits and electrical apparatus and to any parts of the structure of any premises in which it is installed which might affect its safe use.

(2) Any person who connects any installation pipework to a primary meter shall, in any case where equipotential bonding may be necessary, inform the responsible person that such bonding should be carried out by a competent person.

ACOP 18(1)–(2)

200 The location and routing of installation pipework should take into account the potential risk (eg of corrosion damage) posed by the other building services, equipment and features specified in regulation 18(1), for instance by providing adequate separation. Pipework installation and equipotential bonding should be carried out to the appropriate standard.

Guidance 18(1)–(2)

201 Main equipotential bonding (MEB, sometimes known as 'electrical cross bonding') is the connection between the consumer earth point and the gas installation pipe. The purpose is to create a zone (eg within a dwelling), including the area occupied by the gas installation pipework, within which acceptable voltage differences are maintained to avoid the risk of electric shock.

202 The person who installs a section of pipework which connects with the primary meter or emergency control, whether or not the meter or control has yet been fitted, must inform the responsible person for the premises (builder, owner or occupier) of the possible need for MEB where such a requirement did not exist before the work was undertaken. Such bonding should be carried out by a competent person (see regulation 3, Qualification and supervision, for details on training and competence). The advice should be in writing. Although the regulation applies only when new systems are installed and existing ones are modified, similar action needs to be taken if an engineer notices an apparent defect in bonding in other circumstances, eg during maintenance checks (this applies to both main or supplementary equipotential bonding – see also paragraph 206).

203 Main equipotential bonding is most commonly required where:

(a) gas pipework is installed in new premises; or
(b) gas pipework is first installed in existing premises.

Guidance 18(1)–(2)

204 The regulation is directed at the 600 mm (approximate) length of installation pipework at the outlet of the domestic meter installation which is the recommended location of any main equipotential bond in the appropriate standard.

205 The requirements for main equipotential bonding are more strict for certain types of electrical supply to premises (eg those supplies from protective multiple earth (PME) systems – most new electrical supplies will be from such systems). A gas engineer may not be competent to make the necessary judgement, in which case the responsible person needs to be informed of this fact and any further action left to them.

206 Regulation 18(2) does not apply to the installation of a meter. In addition to main bonding, supplementary equipotential pipework bonding may be necessary in locations of increased electric shock risk, eg bathrooms. In such cases, a competent electrician should be consulted.

207 In many commercial and other large sites where gas and electric meters may be remotely located, the bonding is not always possible within a 600 mm distance. A competent electrician needs to consider what action is necessary in these cases.

Regulation 19 Enclosed pipes

Summary of regulation 19

This regulation specifies the restrictions and protective measures for pipes passing through solid walls and floors, cavity walls and building foundations. Conditions are stipulated whereby pipework associated with 'living flame effect' fires may be run in a wall cavity. Ducts and voids accommodating installation pipework must be adequately ventilated.

Regulation 19(1)

(1) No person shall install any part of any installation pipework in a wall or a floor or standing of solid construction unless it is so constructed and installed as to be protected against failure caused by the movement of the wall, the floor or the standing as the case may be.

Guidance 19(1)

208 See the appropriate standard for suitable installation methods (see the *Legislative, normative and informative document list* at www.gassaferegister.co.uk).

Regulation 19(2)–(6)

(2) No person shall install any installation pipework so as to pass through a wall or a floor or standing of solid construction (whether or not it contains any cavity) from one side to the other unless –

(a) any part of the pipe within such wall, floor or standing as the case may be takes the shortest practicable route; and
(b) adequate means are provided to prevent, so far as is reasonably practicable, any escape of gas from the pipework passing through the wall, floor or standing from entering any cavity in the wall, floor or standing.

(3) No person shall, subject to paragraph 4, install any part of any installation pipework in the cavity of a cavity wall unless the pipe is to pass through the wall from one side to the other.

(4) Paragraph (3) shall not apply to the installation of installation pipework connected to a living flame effect gas fire provided that the pipework in the cavity is as short as is reasonably practicable, is enclosed in a gas tight sleeve and sealed at the joint at which the pipework enters the fire; and in this paragraph a "living flame effect gas fire" means a gas fire –

Regulation 19(2)–(6)

(a) designed to simulate the effect of a solid fuel fire;
(b) designed to operate with a fanned flue system; and
(c) installed within the inner leaf of a cavity wall.

(5) No person shall install any installation pipework or any service pipework under the foundations of a building or in the ground under the base of a wall or footings unless adequate steps are taken to prevent damage to the installation pipework or service pipework in the event of the movement of those structures or the ground.

(6) Where any installation pipework is not itself contained in a ventilated duct, no person shall install any installation pipework in any shaft, duct or void which is not adequately ventilated.

Guidance 19(2)–(6)

209 Regulation 19(2)(b) addresses the potential risk of gas leaking from pipework within a cavity in a wall, floor or standing. Such leakage may be difficult to detect and readily lead to the accumulation of an explosive gas/air mixture in the cavity, presenting a considerable hazard to building occupants and others. The protective measures required may include enclosing that part of the pipe which passes through the wall, floor or standing in a gas-tight sleeve which itself is ventilated to a safe position, preferably to open air. This approach also provides some protection against possible mechanical damage to pipework, eg arising from structural movement (see also regulation 19(5)). Any gap between the pipe and sleeve should be sealed with fire-retardant flexible sealant (see standards in the *Legislative, normative and informative document list*), but at one end only to ensure any leak cannot accumulate in the gap space.

210 Shafts, ducts and voids used for accommodation of gas pipework should comply with the appropriate standard. Measures to comply with regulation 19, eg void/duct ventilation arrangements, should not impair any provisions for fire/smoke separation in a building – see regulation 20.

211 When considering LPG pipework, because LPG vapour is heavier than air, cellars, basements or unventilated suspended floors are all considered to be voids. Further guidance on the design, installation and testing of LPG piping systems and maintenance of LPG installations is given in the UKLPG Codes of Practice (see www.uklpg.org).

212 The exception for 'living flame effect' fires in regulation 19(4) applies only to those fires of this type which operate with a fanned flue system (ie where removal of combustion products depends on a fan rather than natural draught through a flue/chimney) and are installed within the inner leaf of a cavity wall. Although pipework connected to these fires may be 'hidden' by routing within a cavity, without the restriction in regulation 19(3), the length of pipework in the cavity must still be kept as short as is reasonably practicable, ie by taking the most direct route consistent with requirements for the particular installation. The installation of living flame effect fires and associated flue systems, as covered by regulation 19(4), normally involves work which may affect the structural integrity of a building; requirements of Building Regulations should be met, as appropriate.

Regulation 20 Protection of buildings

Summary of regulation 20

This regulation requires that installation pipework is installed to avoid impairing the structure or fire resistance of a building.

Regulation 20

No person shall install any installation pipework in a way which would impair the structure of a building or impair the fire resistance of any part of its structure.

Regulation 21 Clogging precautions

Summary of regulation 21

This regulation requires a receptor to be fitted to installation pipework where liquid or solid deposits may occur, eg from 'wet gas'.

Regulation 21

No person shall install any installation pipework in which deposition of liquid or solid matter is likely to occur unless a suitable vessel for the reception of any deposit which may form is fixed to the pipe in a conspicuous and readily accessible position and safe means are provided for the removal of the deposit.

Guidance 21

213 Natural gas and LPG are 'dry' gases and do not normally require clogging precautions. Engineers working with gases which might cause solid or liquid deposits (eg gas produced from landfill or by anaerobic digestion in certain circumstances) should refer to the appropriate standard.

Regulation 22 Testing and purging of pipes

Summary of regulation 22

This regulation specifies the requirements for gastightness testing after work has been done on installation pipework, and for purging/sealing of such pipework both in cases where gas is being supplied to the premises where it is installed, and where gas is not being so supplied.

Regulation 22(1)

(1) Where a person carries out work in relation to any installation pipework which might affect the gastightness of any part of it, he shall immediately thereafter ensure that –

(a) that part is adequately tested to verify that it is gastight and examined to verify that it has been installed in accordance with these Regulations; and

(b) after such testing and examination, any necessary protective coating is applied to the joints of that part.

ACOP 22(1)

214 All joints affected by the portion of work done should be visually inspected to ensure they have been correctly made, as part of the gastightness test.

215 Joints should be tested before being painted or otherwise protected against corrosion.

Guidance 22(1)

216 Paint or corrosion protection applied to a joint before it has been tested could provide a temporary seal and falsify the test result.

Guidance 22(1)

217 Gastightness requirements are set out in the appropriate standard (see the *Legislative, normative and informative document list*).

Regulation 22(2)

(2) Where gas is being supplied to any premises in which any installation pipework is installed and a person carries out work in relation to the pipework, he shall also ensure that –

(a) immediately after complying with the provisions of sub-paragraphs (a) and (b) of paragraph (1) above, purging is carried out throughout all installation pipework through which gas can then flow so as to remove safely all air and gas other than the gas to be supplied;
(b) immediately after such purging, if the pipework is not to be put into immediate use, it is sealed off at every outlet with the appropriate fitting;
(c) if such purging has been carried out through a loosened connection, the connection is retested for gastightness after it has been retightened; and
(d) every seal fitted after such purging is tested for gastightness.

ACOP 22(2)

218 Purging should be carried out in accordance with the appropriate standard and should be undertaken after any work which breaches the installation pipework integrity, not just when gas is first supplied. Requirements in relation to an 'appropriate fitting' are given under the definition in regulation 2(1).

Regulation 22(3)

(3) Where gas is not being supplied to any premises in which any installation pipework is installed –

(a) no person shall permit gas to pass into the installation pipework unless he has caused such purging, testing and other work as is specified in sub-paragraphs (a) to (d) of paragraph (2) above to be carried out;
(b) a person who provides a gas supply to those premises shall, unless he complies with sub-paragraph (a) above, ensure that the supply is sealed off with an appropriate fitting.

Guidance 22(3)

219 For the relevant standards see the *Legislative, normative and informative document list*.

220 This regulation applies when a premises is first supplied with gas. It does not apply where gas has been temporarily cut off, eg where an emergency control has been shut off, or an over-/under-pressure valve has operated. Neither does it apply to situations merely involving a change in gas supplier to a premises.

221 In some cases, a gas supply may be provided some time before installation work is done, or vice versa. Where this occurs, the supply or installation pipework should be properly capped so that no third party, eg the occupant, can easily interfere with it. The person finally connecting the supply to the installation is responsible for testing and purging the pipe system in accordance with regulation 22(2).

Regulation 23 Marking of pipes

Summary of regulation 23

This regulation requires installation pipework (other than in premises or part of premises used only as a dwelling or living accommodation) to be marked, eg colour coded, in any position accessible to inspection to identify that it is carrying gas.

Regulation 23(1)

(1) Any person installing, elsewhere than in any premises or part of premises used only as a dwelling or for living accommodation, a part of any installation pipework which is accessible to inspection, shall permanently mark that part in such a manner that it is readily recognisable as part of a pipe for conveying gas.

Guidance 23(1)

222 This regulation does not apply to domestic premises or living accommodation. In all other locations, eg commercial premises such as offices, gas pipes accessible to inspection should be colour coded and/or marked in accordance with the appropriate standard.

Regulation 23(2)

(2) The responsible person for the premises in which any such part is situated shall ensure that the part continues to be so recognisable so long as it is used for conveying gas.

Regulation 24 Large consumers

Summary of regulation 24

This regulation requires that a valve must be fitted in certain installation pipework and a system diagram provided (eg for use by emergency services) where service pipe/pipework exceeding specified sizes feeds certain buildings or floor areas.

Regulation 24(1)–(3)

(1) Where the service pipe to any building having two or more floors to which gas is supplied or (whether or not it has more than one floor) a floor having areas with a separate supply of gas has an internal diameter of 50 mm or more, no person shall install any incoming installation pipework supplying gas to any of those floors or areas, as the case may be, unless –

- *(a) a valve is installed in the pipe in a conspicuous and readily accessible position; and*
- *(b) a line diagram in permanent form is attached to the building in a readily accessible position as near as practicable to –*
  - *(i) the primary meter or where there is no primary meter, the emergency control, or*
  - *(ii) the gas storage vessel, indicating the position of all installation pipework of internal diameter of 25 mm or more, meters, emergency controls, valves and pressure test points of the gas supply systems in the building.*

(2) Paragraph (1) above shall apply to service pipework as it applies to a service pipe except that reference therein to "50 mm or more" is to be reference to "30 mm or more".

(3) In paragraph (1)(b) above "pressure test point" means a gas fitting to which a pressure gauge can be connected.

Guidance 24(1)–(3)

223 This regulation applies where a service pipe has an internal diameter of 50 mm or more, or service pipework has an internal diameter of 30 mm or more, and divides to supply more than one floor or separate areas on one floor. This situation is most often encountered in industrial and commercial premises.

224 Where, to comply with regulation 24(1)(a), an isolation valve needs to be installed in the service pipe, the gas transporter must be notified and arrangements made, eg for isolation and degassing as necessary, to ensure this can be done safely.

225 The system diagram should be prominently displayed at or near to the primary meter, or if there is no such meter, the emergency control. Where gas is supplied from a storage vessel, the notice may alternatively be displayed at or near to that vessel. Depending on the size of the premises, other copies may be held elsewhere, eg the security gate or reception area.

226 The purpose of the system diagram is to allow anyone, especially the emergency services, to identify and isolate part of the gas system if necessary. Pipes of less than 25 mm do not need to be shown on the diagram but it is recommended that additional information to that specified in regulation 24(1)(a) be included, where significant to the safety of a particular installation, eg this might include purge connections. The diagram needs to be updated as necessary.

PART E Gas appliances

Summary of regulations 25 and 26

Regulations 25 and 26 impose a duty on any person installing a gas appliance to ensure:

(a) it is safe for use;
(b) it is not left connected to the gas supply unless it can be used safely;
(c) it complies with other relevant safety requirements (eg gas appliances safety legislation); and
(d) that any second-hand appliance is in a safe condition for further use.

Any work on an appliance must maintain safety. The appliance must be examined after work has been done, and any defect notified to the owner/user.

Regulation 25 Interpretation of Part E

Regulation 25

In this Part –

"flue pipe" means a pipe forming a flue but does not include a pipe built as a lining into either a chimney or a gas appliance ventilation duct;

"operating pressure", in relation to a gas appliance, means the pressure of gas at which it is designed to operate.

Regulation 26 Gas appliances – safety precautions

Regulation 26(1)

(1) No person shall install a gas appliance unless it can be used without constituting a danger to any person.

ACOP 26(1)

227 Gas engineers should ensure that any appliance they install, or flue to which they connect an appliance, is safe for use. Requirements in regulations 26(9) and 33 should be met as applicable, and reference made to appropriate standards.

228 Appliances should be installed in accordance with manufacturers' instructions, including any manufacturers' instructions on flues to which the appliances can be safely connected.

Guidance 26(1)

229 Appendix 3 provides details of appropriate tests and examinations that should be carried out. A list of the relevant standards can be found in the *Legislative, normative and informative document list*.

230 This regulation imposes an important and wide-ranging duty on the engineer to ensure that nothing about an appliance itself, the manner in which it is installed, any associated fitting or flue, or other factor (eg ventilation) will cause danger when the appliance is taken into use. This 'all-embracing' duty interfaces with other parts of GSIUR dealing with more specific safety aspects of appliance installation, and

Guidance 26(1)

requirements under regulation 26(9) for examination/testing after any work on an appliance, including installation.

231 In addition to the duty under regulation 26(1), regulation 30 places restrictions/prohibitions on the installation of certain appliances. The installation of an open-flued appliance in the same room (this includes a through lounge) as a solid-fuel open fire, or an appliance with a fan-assisted flue, is not recommended because it may have an adverse effect on the performance of an existing flue (see the *Legislative, normative and informative document list*).

Regulation 26(2)

(2) No person shall connect a flued domestic gas appliance to the gas supply system except by a permanently fixed rigid pipe.

Guidance 26(2)

232 Flexible connections should not be used as this could allow consumers to move the appliance away from the flue. See also regulation 5(2) concerning construction materials for pipes and pipe fittings.

Regulation 26(3)

(3) No person shall install a used gas appliance without verifying that it is in a safe condition for further use.

Guidance 26(3)

233 The physical condition of any used appliance should be checked before installation. This is necessary as suitable inspection might not be possible after an appliance has been installed.

234 In addition to second-hand appliances, this regulation also applies when an appliance is moved from one location to another, even within the same room.

Regulation 26(4)

(4) No person shall install a gas appliance which does not comply with any enactment imposing a prohibition or restriction on the supply of such an appliance on grounds of safety.

Guidance 26(4)

235 New appliances should conform to the Gas Appliances (Safety) Regulations 1995 (see summary of legislation at www.hse.gov.uk/gas/domestic/).

Regulation 26(5)

(5) No person carrying out the installation of a gas appliance shall leave it connected to the gas supply unless –

(a) the appliance can be used safely; or
(b) the appliance is sealed off from the gas supply with an appropriate fitting.

Guidance 26(5)

236 An appliance should be properly commissioned and checked for safe operation immediately after it has been installed and connected to the gas supply (see also regulation 33). If this cannot be done, the appliance should be sealed off from the gas supply by capping off the installation pipework serving the appliance, using an appropriate fitting. For requirements concerning an 'appropriate fitting' see the definition in regulation 2(1) and associated guidance.

Regulation 26(6)

(6) No person shall install a gas appliance without there being at the inlet to it means of shutting off the supply of gas to the appliance unless the provision of such means is not reasonably practicable.

Guidance 26(6)

237 The means of shutting off the gas supply to an appliance may comprise either a conventional isolation valve or other effective means, eg a self-sealing plug-in connector (commonly used on cooker installations) or a screw-down restrictor elbow (such as used on gas fire installations). A shut-off device at the inlet to a gas appliance is regarded as part of 'installation pipework' for the purpose of GSIUR (see definition in regulation 2(1)).

Guidance 26(6)

238 The shut-off device required under this regulation assists tightness testing of installation pipework and allows an appliance to be isolated/disconnected in the case of an emergency, eg if the appliance develops a gas leak or otherwise becomes dangerous. Although the regulation recognises that provision of a shut-off device at the inlet of an appliance might occasionally be difficult, there will be few, if any, situations where this is not reasonably practicable.

Regulation 26(7)–(10)

(7) No person shall carry out any work in relation to a gas appliance which bears an indication that it conforms to a type approved by any person as complying with safety standards in such a manner that the appliance ceases to comply with those standards.

(8) No person carrying out work in relation to a gas appliance which bears an indication that it so conforms shall remove or deface the indication.

(9) Where a person performs work on a gas appliance he shall immediately thereafter examine –

(a) the effectiveness of any flue;
(b) the supply of combustion air;
(c) subject to sub-paragraph (ca), its operating pressure or heat input or, where necessary, both;
(ca) if it is not reasonably practicable to examine its operating pressure or heat input (or, where necessary, both), its combustion performance;[1]
(d) its operation so as to ensure its safe functioning,

and forthwith take all reasonably practicable steps to notify any defect to the responsible person and, where different, the owner of the premises in which the appliance is situated or, where neither is reasonably practicable, in the case of an appliance supplied with liquefied petroleum gas, the supplier of gas to the appliance, or, in any other case, the transporter.

(10) Paragraph 9 shall not apply in respect of –

(a) the direct disconnection of the gas supply of a gas appliance; or
(b) the purging of gas or air from an appliance or its associated pipework or fittings in any case where that purging does not adversely affect the safety of that appliance, pipe or fitting.

ACOP 26(7)–(10)

239 After performing work on an installed appliance, engineers should carry out the necessary checks and tests to ensure that the appliance, and any associated flue, is safe for use (see regulation 27(1) for information on flues).

240 Reference should be made to manufacturers' instructions and appropriate standards.

Guidance 26(7)–(10)

241 Appendix 3 provides details of appropriate tests and examinations that should be carried out. A list of the relevant standards can be found in the *Legislative, normative and informative document list*.

242 Nothing should be done to a gas appliance which will result in it no longer complying with the standard(s) to which it purports to comply, ie as required by the conformity (eg 'CE') marking on the appliance. Any repair to a gas appliance, for instance involving use of refurbished or reconditioned parts, should be done in a way which ensures safety standards are maintained. Engineers should refer to manufacturers' specifications/instructions and appropriate standards.

1 Regulation 26(9)(c) was amended by the Gas Safety (Installation and Use) (Amendment) Regulations 2018.

Guidance 26(7)–(10)

243 Regulation 26(9) requires the operating (ie gas) pressure and/or heat input of the appliance to be checked. The engineer should carry out the most appropriate test(s) for the appliance. This information is usually shown in manufacturers' installation instructions, or on the appliance data badge. In certain cases, eg when appliance burners are modified or replaced, such as for conversion from LPG to natural gas, it may be necessary to check both operating pressure and heat input.

244 Regulation 26(9) applies only to work on gas appliances; it does not cover work on certain other gas fittings, eg meters and installation pipework (that might be carried out remote from an appliance). See definitions of 'gas appliance' and 'gas fittings' in regulation 2(1). Appropriate checks and tests are still required to ensure safety after any work on such fittings, eg see regulations 6(6), 12(6) and 22(1). See also exceptions in regulation 2(6)(c) for work on certain gas appliance control devices primarily intended for use by the consumer.

245 Regulation 26(9)(ca), applies where there is no meter or meter display to directly measure the heat input and it is not possible to measure the operating pressure.

246 Under regulation 26(10), the tests/examinations in regulation 26(9) are not required after certain work on appliances. The exception applies, for instance, to the direct disconnection of an appliance, ie where the appliance is itself detached from the gas supply, such as at the appliance connector or a nearby pipework connection. (Indirect disconnection, eg resulting from removal of another fitting upstream, such as a meter, is separately excluded as this is not regarded as work on a gas appliance.)

247 Regulation 26(9) also does not apply to purging of air or gas from an appliance, eg in the process of resuming a gas supply after an interruption, but this is conditional on the purging operation itself having no adverse effect on the safety of any gas fitting. (This would not preclude the possibility of an appliance defect present before, and completely unrelated to, the purging activity still remaining after purging has been completed; such defects would need to be detected and addressed separately – see regulations 35 and 36.)

248 To establish whether the safety of any gas fitting might be affected by purging, a competent person should make an assessment before the operation is carried out. In any case of doubt or where safety is likely to be affected, the tests/examinations in regulation 26(9) should be carried out after purging has been completed, and remedial action taken to make appliances/fittings safe, as necessary.

249 It should be stressed that the exceptions in regulation 26(10) apply only to those activities specifically mentioned and if other work is carried out in association with these activities, eg any adjustment of appliances after purging, the requirements in regulation 26(9) again need to be met.

250 Engineers need to take account of manufacturers' instructions and any safety warnings given by them, especially any recall-type information about safety problems encountered with appliances some time after the particular model was first introduced onto the market.

251 If any safety defect is found as a result of the tests/examinations under regulation 26(9), the responsible person, and (if appropriate) the landlord for the premises, should be notified as soon as possible. If neither is available, the gas supplier or transporter should be notified (as appropriate) – see also regulation 34 concerning unsafe appliances. Further information is available in *The Gas Industry Unsafe Situations Procedure*.

Regulation 27 Flues

Summary of regulation 27

This regulation requires that any flue must be suitable and in a proper condition for the appliance it serves to operate safely. It also requires that any power-operated flue system should prevent operation of the appliance if the draught fails. Requirements to enable inspection of, and to prevent spillage or leakage of combustion products from, certain flues are specified. All flues must be installed in a safe position.

Regulation 27(1)

(1) No person shall install a gas appliance to any flue unless the flue is suitable and in a proper condition for the safe operation of the appliance.

ACOP 27(1)

252 Whenever an appliance is installed to a flue, the installer should ensure that the flue is adequate, suitable and effective for use with the appliance which it will serve. An effective flue should be complete and continuous throughout its length and safely carry the products of combustion from a gas appliance to the external air.

253 This requirement applies however the connection is made (eg it covers equally where an appliance is fitted to an existing flue, and where a flue is fitted to an existing appliance). In each case, the necessary tests and examinations should be carried out both before and after the appliance has been fitted. Reference should be made to appropriate standards and the manufacturers' instructions.

Guidance 27(1)

254 The requirements of Building Regulations concerning flues and chimneys should be met, where applicable. Fitting a suitable terminal to any flue/chimney outlet is recommended to help products of combustion to escape, minimise downdraught and prevent entry of material (eg bird nests) that might cause blockage. In certain areas, eg where squirrels inhabit or birds such as jackdaws are known to roost, purpose-designed protection is essential to avoid such blockage (see paragraph 143).

255 It should not be assumed that previous satisfactory operation means that a flue will continue to be safe for use with another appliance, even of the same type. Appropriate tests and examinations should always be carried out – see Appendix 3. Guidance on types of appliance which require a flue to be fitted is given in the appropriate standards (see the *Legislative, normative and informative document list*).

Regulation 27(2)–(3)

(2) No person shall install a flue pipe so that it enters a brick or masonry chimney in such a way that the seal between the flue pipe and chimney cannot be inspected.

(3) No person shall connect a gas appliance to a flue which is surrounded by an enclosure unless that enclosure is so sealed that any spillage of products of combustion cannot pass from the enclosure to any room or internal space other than the room or internal space in which the appliance is installed.

Guidance 27(2)–(3)

256 Where a false chimney breast or decorative canopy is fitted, it should be sealed from other rooms in the premises. Reference should be made to appropriate standards.

Regulation 27(4)

(4) No person shall install a power operated flue system for a gas appliance unless it safely prevents the operation of the appliance if the draught fails.

Guidance 27(4)

257 An interlock should be provided to cut off the gas supply if the draught falls below a preset minimum standard for safe operation of the appliance, and prevent the gas supply being re-established unless safe to do so. The advice of the appliance/flue system designer/manufacturer should be sought, as necessary, in respect of interlock design and flue performance monitoring requirements. Refer to appropriate standards, eg concerning provision of manual reset facilities and interlocks with flame-proving devices. See also regulation 32 and associated ACOP and guidance concerning flue dampers.

Regulation 27(5)

(5) No person shall install a flue other than in a safe position.

ACOP 27(5)

258 A flue (including any terminal) should be installed in a position which ensures that it will operate effectively and that products of combustion will safely disperse and not present a hazard to any person, whether in the premises where the associated appliance is installed (eg by being located a safe distance from vents and openable windows), or in adjoining/neighbouring premises. The location needs to take into account any possible developments in neighbouring property, eg building extensions.

259 Any flue should therefore be sited to discharge at a safe distance from any boundary with adjoining premises (see also regulation 8). Refer to requirements in Building Regulations, appropriate standards and manufacturers' instructions.

260 A flue should not be sited in such a position where it cannot be subsequently examined.

Regulation 28 Access

Summary of regulation 28

This regulation requires that a gas appliance must be installed in a position readily accessible for operation, inspection and maintenance.

Regulation 28

(1) No person shall install a gas appliance except in such a manner that it is readily accessible for operation, inspection and maintenance.

Guidance 28

261 Minimum clearance distances for operation, inspection and maintenance purposes are normally specified in manufacturers' instructions for the installation of appliances.

Regulation 29 Manufacturer's instructions

Summary of regulation 29

This regulation requires that the installer of a gas appliance must leave the manufacturer's instructions for the appliance for use by the owner or occupier of the premises where the appliance is installed.

Regulation 29

Any person who installs a gas appliance shall leave for the use of the owner or occupier of the premises in which the appliance is installed all instructions provided by the manufacturer accompanying the appliance.

Guidance 29

262 Requirements concerning manufacturers' instructions to accompany appliances are covered by the Gas Appliances (Safety) Regulations 1995 (see summary of legislation at www.hse.gov.uk/gas/domestic/).

Regulation 30 Room-sealed appliances

Summary of regulation 30

This regulation prohibits the installation of certain gas appliances in specified rooms, unless the appliance is room-sealed. In other specified locations, certain appliances must be room-sealed or fitted with a device to cause shutdown before a dangerous quantity of combustion products can build up in the room concerned. A general prohibition is placed on the installation of any instantaneous water heater, unless it is room-sealed or fitted with such a device.

Regulation 30(1)–(4)

(1) No person shall install a gas appliance in a room used or intended to be used as a bathroom or a shower room unless it is a room-sealed appliance.

(2) No person shall install a gas fire, other gas space heater or a gas water heater of more than 14 kilowatt gross heat input in a room used or intended to be used as sleeping accommodation unless the appliance is a room-sealed appliance.

(3) No person shall install a gas fire, other gas space heater or a gas water heater of 14 kilowatt gross heat input or less in a room used or intended to be used as sleeping accommodation and no person shall install an instantaneous water heater unless (in each case) –

(a) it is a room-sealed appliance; or
(b) it incorporates a safety control designed to shut down the appliance before there is a build up of a dangerous quantity of the products of combustion in the room concerned.

(4) The references in paragraphs (1) to (3) to a room used or intended to be used for the purpose therein referred to includes a reference to –

(a) a cupboard or compartment within such a room; or
(b) a cupboard, compartment or space adjacent to such a room if there is an air vent from the cupboard, compartment or space into such a room.

Guidance 30(1)–(4)

263 Under regulation 30(2), non-room-sealed appliances used for water heating, space heating and central heating of 14 kW or more gross heat input should not be installed in accommodation designed for sleeping in, such as bedrooms, bed-sitting rooms and the sleeping areas of caravans. However, non-room-sealed heating appliances of less than 14 kW gross heat input may be fitted in these rooms/areas provided they incorporate a device which turns off the gas supply before a dangerous level of fumes can build up (regulation 30(3)). An instantaneous water heater that is not room-sealed may not be installed in any room unless it incorporates such a device.

264 The prohibitions in regulation 30 each extend to any cupboard or compartment which is accessed from the accommodation or room concerned (see regulation 30(4)). They also extend to any cupboard, compartment (eg cubicle) or space adjoining the accommodation or room concerned, where ventilation for safe operation of the appliance is provided via the accommodation or room concerned. Spaces adjoining the accommodation or room concerned where there is no air vent from the space into the room, ie which are provided with ventilation for safe operation of an appliance from another source, are not included in the prohibition.

Guidance 30(1)–(4)

265 The rooms subject to the restrictions in regulation 30 are not only those actually being used as bathrooms, shower rooms or sleeping accommodation, but also those known to be intended for such use, at the time when the appliance is installed. The references to 'bathroom' and 'shower room' in this regulation include any room not purpose-built as such, but containing a functioning bath or shower, and the restrictions on appliance installation apply accordingly.

266 The prohibitions in regulation 30 also extend, in certain situations, to appliances fitted in rooms converted, or intended to be converted, by landlords into sleeping accommodation (see regulation 36(11)–(12)).

Regulation 31 Suspended appliances

Summary of regulation 31

This regulation prohibits the installation of suspended appliances unless the installation pipework is capable of supporting the weight imposed and the appliance is designed to be so supported.

Regulation 31

No person shall install a suspended gas appliance unless the installation pipework to which it is connected is so constructed and installed as to be capable of safely supporting the weight imposed on it and the appliance is designed to be so supported.

Guidance 31

267 The installation of appliances by suspension from installation pipework should be avoided, wherever practicable. Where this approach has to be considered, it should be ensured that the appliance is suitable for installation in this way and that the pipework is properly supported and capable of safely withstanding the weight imposed. Reference should be made to the appliance manufacturer's instructions.

Regulation 32 Flue dampers

Summary of regulation 32

This regulation describes the requirements for interlocking automatic flue dampers and their inspection. Installation of a manual flue damper on a domestic appliance is prohibited, and where an appliance is installed to an existing flue incorporating a manual flue damper, the damper must be permanently fixed in the open position.

Regulation 32(1)–(3)

(1) Any person who installs an automatic damper to serve a gas appliance shall –

(a) ensure that the damper is so interlocked with the gas supply to the burner that burner operation is prevented in the event of failure of the damper when not in the open position; and
(b) immediately after installation examine the appliance and the damper to verify that they can be used together safely without constituting a danger to any person.

(2) No person shall install a manually operated damper to serve a domestic gas appliance.

(3) No person shall install a domestic gas appliance to a flue which incorporates a manually operated damper unless the damper is permanently fixed in the open position.

ACOP 32(1)–(3)

268 Reference should be made to appropriate standards and appliance manufacturers' instructions.

Guidance 32(1)–(3)

269 An automatic damper may be used with a gas appliance subject to conditions in regulation 32(1) being satisfied. A damper system meeting these requirements will both prevent gas being supplied to the burner (unless the damper is in the fully open position) and provide for adequate pre-ignition purging and burner post-purging on flame failure. Purging is particularly important for enclosed appliances where ignition of an unburned gas/air mixture in the combustion system has the potential to cause a serious explosion. See also regulation 27(4) concerning power-operated flue systems.

270 Fitting a manually operated damper system to serve a domestic gas appliance is prohibited under regulation 32(1) because, although it is possible to interlock the damper with the gas supply, an assured means of purging is not generally possible with such a system (see paragraph 269).

271 Regulation 32(3) makes provision for an appliance to be fitted to an existing flue incorporating a manual damper, but only where the damper is permanently fixed in the open position. A manual damper may, for instance, often be present in a hearth used for burning coal or other solid fuel. Any such damper should be removed (this is often necessary for 'flame-effect' fires), or permanently secured in the fully open position, to protect against the possibility of interference by a third party (such as the occupant) or the damper falling into a closed position, eg as a result of corrosion damage.

Regulation 33 Testing of appliances

Summary of regulation 33

This regulation describes the requirements for testing gastightness and examining appliances, flues, ventilation etc when appliances are installed. It details requirements for when appliances are installed at premises both at a time when gas is being supplied and when gas is not being supplied. It also details action where adjustments are necessary.

Regulation 33(1)

(1) Where a person installs a gas appliance at a time when gas is being supplied to the premises in which the appliance is installed, he shall immediately thereafter test its connection to the installation pipework to verify that it is gastight and examine the appliance and the gas fittings and other works for the supply of gas and any flue or means of ventilation to be used in connection with the appliance for the purpose of ascertaining whether –

(a) the appliance has been installed in accordance with these Regulations;
(b) the operating pressure is as recommended by the manufacturer;
(c) the appliance has been installed with due regard to any manufacturer's instructions provided to accompany the appliance; and
(d) all gas safety controls are in proper working order.

ACOP 33(1)

272 Testing procedures should be in accordance with appropriate standards and manufacturers' instructions.

Guidance 33(1)

273 A list of appropriate standards can be found in the *Legislative, normative and informative document list.*

Regulation 33(2)

(2) Where a person carries out such testing and examination in relation to a gas appliance and adjustments are necessary to ensure compliance with the requirements specified in sub-paragraphs (a) to (d) of paragraph (1) above, he shall either carry out those adjustments or disconnect the appliance from the gas supply or seal off the appliance from the gas supply with an appropriate fitting.

ACOP 33(2)

274 Once the commissioning of an appliance is started, it should be either completed in full, leaving the appliance in a safe working condition, or the appliance should be disconnected or the gas supply to the appliance sealed off with an appropriate fitting, until tests and examinations can be fully completed at a later date. See also definition of 'appropriate fitting' in regulation 2(1).

Guidance 33(2)

275 In addition to disconnection/sealing off from the gas supply as in paragraph 274, a label needs to be attached to any appliance which has not been fully commissioned indicating this to be the case. The label should state that the appliance should not be reconnected to the gas supply, or the appropriate fitting sealing off the gas supply to the appliance removed, until the procedure has been safely completed. Where an appliance is disconnected, the outlet of every pipe to which it was connected should be sealed off in accordance with regulation 6(3).

Regulation 33(3)

(3) Where gas is not being supplied to any premises in which any gas appliance is installed –

(a) no person shall subsequently permit gas to pass into the appliance unless he has caused such testing, examination and adjustment as is specified in paragraphs (1) and (2) above to be carried out; and

(b) a person who subsequently provides a gas supply to those premises shall, unless he complies with sub-paragraph (a) above, ensure that the appliance is sealed off from the gas supply with an appropriate fitting.

ACOP 33(3)

276 A person who allows gas to flow into any appliance should ensure that gas fittings are purged and tested for safety, or the supply is sealed off with a fitting to the appropriate standard (see also definition of 'appropriate fitting' in regulation 2(1)).

Guidance 33(3)

277 This regulation applies when a premises is first supplied with gas. In some cases, a gas supply may be provided some time before any installation work is done, or vice versa. Where this occurs, the supply or appliance should be sealed with an appropriate fitting so that no third party, eg the occupant, can easily interfere with it. The person finally connecting the supply to the appliance is responsible for testing and purging the system and appliance in accordance with regulations 22 and 33(1)–(2).

278 This regulation does not apply where gas has been temporarily cut off from an appliance, eg where the emergency control has been shut off or an over/under-pressure valve has operated. Neither does it cover situations merely involving a change in gas supplier to a premises.

Regulation 34 Use of appliances

Summary of regulation 34

This regulation imposes a duty on the responsible person for any premises (for instance, the occupier or owner (eg landlord) of the premises) not to use or permit the use of any unsafe appliance. People carrying out specified work (eg on service pipes or gas fittings) must report any appliance they suspect is dangerous to the responsible person for the premises, or where this person is not available, to the gas supplier or transporter, as appropriate.

Regulation 34(1)–(2)

(1) The responsible person for any premises shall not use a gas appliance or permit a gas appliance to be used if at any time he knows or has reason to suspect that it cannot be used without constituting a danger to any person.

Regulation 34(1)–(2)

(2) For the purposes of paragraph (1) above, the responsible person means the occupier of the premises, the owner of the premises and any person with authority for the time being to take appropriate action in relation to any gas fitting therein.

ACOP 34(1)–(2)

279 A responsible person must not use or allow the use of any appliance which it is known or suspected could constitute a danger to any person, and in particular danger of fire/explosion arising from gas leakage or carbon monoxide poisoning arising from inadequate flueing arrangements or fixed ventilation.

280 An unsafe/dangerous appliance in this context means any appliance where both (a) and (b) which follow apply:

- **(a) there is a design, construction, installation, modification, servicing/maintenance deficiency or other fault (eg maladjustment) in the gas appliance, associated flue/ventilation arrangement or a gas fitting/other works for the gas supply to the appliance, which has or is likely to result in:**
  - **(i) incomplete gas combustion; or**
  - **(ii) products of combustion not being safely removed; or**
  - **(iii) insufficient oxygen being available for the occupants of the room/space where the appliance is located; or**
  - **(iv) an accidental gas leakage; or**
  - **(v) other danger, eg of fire;**
- **(b) the resulting leakage, inadequate combustion, inadequate removal of the products of combustion, insufficiency of oxygen or other danger has caused or is likely to cause death or serious injury.**

Guidance 34(1)–(2)

281 Users and other people with control over the use of gas appliances (as specified in regulation 34(2)) must not use or permit use of any gas appliance they know or suspect to be unsafe, eg when they have been told so by a competent gas engineer or emergency service provider. In the case of rented property this duty extends to tenants, landlords and managing agents.

282 The requirement applies only to 'unsafe' appliances (see paragraph 280); it does not cover those which are merely 'sub-standard' and considered in need of minor improvements to bring them up to current standards.

283 General examples of the kinds of fault described in paragraph 280(a) which, if likely to cause death or serious injury, would be regarded as an 'unsafe' appliance, are given in *The Gas Industry Unsafe Situations Procedure*. However, the level of risk (ie whether an appliance is 'dangerous' in a particular case) has to be a matter for judgement by a competent person, based on their knowledge and experience, and taking into account the specific circumstances.

284 In some cases, the danger will be clear, such as:

- (a) where gas is escaping or there is a high risk of this happening (eg open-ended pipes connected to the gas supply);
- (b) gas appliance safety devices (flame protection etc) are inoperative; or
- (c) where there are clear signs of products of combustion being emitted into occupied areas (and there is no evidence to show that the problem has been corrected).

285 See also regulation 37 and associated guidance concerning action in the event of a gas escape.

Guidance 34(1)–(2)

286 If a gas engineer discovers an unsafe gas appliance, the user and, if different, the owner of the appliance (in rented accommodation, the landlord; or managing agent where the landlord is absent or has entrusted the agent to deal with gas safety matters on their behalf) should be informed – in writing – that the appliance is unsafe and could cause death or serious injury and that continued use is an offence.

287 The engineer should seek to persuade the user/owner to allow them to repair or disconnect the appliance but the engineer has no legal power to take such action if the user/owner disagrees. If consent for such action is not obtained, the appliance should be suitably labelled as being unsafe and could cause death or serious injury and that continued use is an offence. With the agreement of the responsible person, the appliance shut-off device (if one is fitted) should be closed. Irrespective of any action taken by the gas engineer, the obligation rests on the responsible person(s) not to use or allow the use of any unsafe gas appliance.

288 In line with *The Gas Industry Unsafe Situations Procedure*, where agreement of the responsible person cannot be obtained for repair or disconnection of a dangerous appliance, it is recommended that the engineer contacts:

(a) in the case of an appliance supplied with LPG, the gas supplier; or
(b) in any other case including natural gas, the Gas Emergency Freephone Number: 0800 111 999.

289 In an emergency situation a public gas transporter has powers to enter property and take action to avert danger to life (and property) under the Gas Safety (Rights of Entry) Regulations 1996. In the case of LPG, a contractual right of entry may exist between the supplier and customer. Useful numbers can be found at www.gassaferegister.co.uk.

290 In addition to requiring reports of specified gas incidents, the Reporting of Injuries, Diseases and Dangerous Occurrences Regulations 2013 (RIDDOR) require a report of certain dangerous fittings to be made to HSE (see RIDDOR regulation 11(2)). A report will generally be required where disconnection of an appliance is considered to be necessary. Further information is given in *The Gas Industry Unsafe Situations Procedure*.

Regulation 34(3)–(4)

(3) Any person engaged in carrying out any work in relation to a gas main, service pipe, service pipework, gas storage vessel or gas fitting who knows or has reason to suspect that any gas appliance cannot be used without constituting a danger to any person shall forthwith take all reasonably practicable steps to inform the responsible person for the premises in which the appliance is situated and, where different, the owner of the appliance or, where neither is reasonably practicable, in the case of an appliance supplied with liquefied petroleum gas, the supplier of gas to the appliance, or, in any other case, the transporter.

(4) In paragraph (3) above, the expression "work" shall be construed as if, in the definition of "work" in regulation 2(1) above, every reference to a gas fitting were a reference to a gas main, service pipe, service pipework, gas storage vessel or gas fitting.

Guidance 34(3)–(4)

291 Regulation 34(3) places a duty on a person engaged in any work specified in regulation 34(4) (eg a gas engineer) who becomes aware of an unsafe/dangerous appliance (see paragraph 280) to take steps to inform – in writing – the responsible person for the premises concerned.

Guidance 34(3)–(4)

292 The responsible person may be the landlord/managing agent of rented accommodation, and (where different) the gas appliance owner. Where the responsible person is not available or readily contactable, the gas supplier or transporter (as appropriate) must be notified to enable any necessary action to be taken to make the appliance safe.

293 Details of the gas supplier/transporter emergency gas service may be found on a notice near the meter, or where there is no meter, the emergency control (see regulations 9 and 15). If there is no notice, or the relevant information is not given, the information may be found in the telephone directory or from the Gas Emergency Telephone Numbers section of *The Gas Industry Unsafe Situations Procedure*.

PART F Maintenance

Regulation 35 Duties of employers and self-employed persons

Summary of regulation 35

This regulation requires an employer or self-employed person to ensure that any gas appliance, flue or installation pipework installed at a place of work they control is maintained in a safe condition.

Regulation 35

It shall be the duty of every employer or self-employed person to ensure that any gas appliance, installation pipework or flue installed at any place of work under his control is maintained in a safe condition so as to prevent risk of injury to any person.

Guidance 35

294 The duty to maintain appliances, flues and installation pipework is linked to other related workplace safety responsibilities, eg under the Health and Safety at Work etc Act 1974 or the Provision and Use of Work Equipment Regulations 1998 (see summary of legislation at www.hse.gov.uk/gas/domestic/). This provision only applies to workplaces falling within the scope of GSIUR (see regulations 2(4)–(5)), but similar duties are imposed by separate safety legislation in other places, eg factories (see paragraphs 79–80).

295 The duty under regulation 35 extends to certain mobile and portable appliances, eg LPG space heaters (see definition of gas appliance in regulation 2(1)). Maintenance of service pipework is not covered by regulation 35, but duties in this respect are applied under the general provisions of the Health and Safety at Work etc Act 1974 and related legislation (see summary of legislation at www.hse.gov.uk/gas/domestic/).

296 Effective maintenance of appliances usually involves an ongoing programme of regular/periodic inspections, together with any necessary remedial work. In the absence of specific manufacturers' instructions, effective maintenance should include as a minimum:

(a) examination of the physical condition and safe functioning of appliance(s), installation pipework, ventilation and any flue for deterioration;
(b) carrying out performance tests; and
(c) taking the necessary remedial action.

Reference should be made to Appendix 3 and appropriate standards; see the *Legislative, normative and informative document list*).

297 In the case of some gas equipment and fittings (eg flexible pipework; certain protective devices etc) where inadequate maintenance could cause failure in a dangerous way, a formal system of planned preventative maintenance might be necessary. This aims to prevent failures occurring while equipment is in use and involves a system of written instructions (eg based on appliance/fitting

Guidance 35

manufacturers' instructions) used to initiate inspection, testing and periodic replacement or refurbishing of components or equipment before they reach the end of their useful life. Further information is given in the HSE publication *Safe use of work equipment. Provision and Use of Work Equipment Regulations 1998. Approved Code of Practice and guidance.*[8]

298 Where the appropriate standard or the manufacturer recommends a limited lifetime for a gas fitting (including plastic pipework), it should be replaced before the end of its limited life, unless it can be shown that continued use will not constitute a hazard. The manufacturer should be consulted to establish whether the lifetime of the gas fitting can be extended, what tests need to be carried out, and the extended lifetime allowed. See also regulations 35–36 and associated guidance, concerning maintenance/safety-check duties.

299 Where non-domestic premises such as public houses or offices are leased as workplaces, employers' duties under regulation 35 may interface with landlords' responsibilities under section 4 of the Health and Safety at Work etc Act 1974, eg for maintenance of gas heating appliances. In these situations, close co-operation and clear allocation of responsibilities are essential to ensure that requirements are fully met and no gaps in safety cover can arise. The landlord and tenant (ie employer/self-employed person) may come to a contractual arrangement in such cases, and such action is recommended to ensure responsibilities are clearly and unambiguously defined.

Regulation 36 Duties of landlords

Summary of regulation 36

This regulation details landlords' duties in respect of gas safety. There are two main duties on landlords: annual safety checks on gas appliances and flues (and that a record is kept and issued, or in certain cases, displayed to tenants) and ongoing maintenance. Although related, these duties are separate and distinct.

In addition landlords must ensure that no gas fitting of a type that would contravene regulation 30 (eg certain instantaneous water heaters) is fitted in any room occupied or to be occupied as sleeping accommodation after the Regulations came into force. This includes any room converted into such accommodation after that time.

Further detailed guidance for landlords on their duties under regulation 36 and how to meet them can be found on the HSE website (www.hse.gov.uk/gas).

Regulation 36(1)–(2)

(1) In this regulation –

"landlord" means –

(a) in England and Wales –
- *(i) where the relevant premises are occupied under a lease, the person for the time being entitled to the reversion expectant on that lease or who, apart from any statutory tenancy, would be entitled to possession of the premises; and*
- *(ii) where the relevant premises are occupied under a licence, the licensor, save that where the licensor is himself a tenant in respect of those premises, it means the person referred to in paragraph (i) above;*

(b) in Scotland, the person for the time being entitled to the landlord's interest under a lease;

Regulation 36(1)–(2)

"lease" means –

- *(a) a lease for a term of less than 7 years; and*
- *(b) a tenancy for a periodic term; and*
- *(c) any statutory tenancy arising out of a lease or tenancy referred to in sub-paragraphs (a) or (b) above,*

and in determining whether a lease is one which falls within sub-paragraph (a) above –

- *(i) in England and Wales, any part of the term which falls before the grant shall be left out of account and the lease shall be treated as a lease for a term commencing with the grant;*
- *(ii) a lease which is determinable at the option of the lessor before the expiration of 7 years from the commencement of the term shall be treated as a lease for a term of less than 7 years;*
- *(iii) a lease (other than a lease to which sub-paragraph (b) above applies) shall not be treated as a lease for a term of less than 7 years if it confers on the lessee an option for renewal for a term which, together with the original term, amounts to 7 years or more; and*
- *(iv) a "lease" does not include a mortgage term;*

"relevant gas fitting" means –

- *(a) any gas appliance (other than an appliance which the tenant is entitled to remove from the relevant premises) or any installation pipework installed in any relevant premises; and*
- *(b) any gas appliance or installation pipework which, directly or indirectly, serves the relevant premises and which either –*
  - *(i) is installed in any part of premises in which the landlord has an estate or interest; or*
  - *(ii) is owned by the landlord or is under his control,*

except that it shall not include any gas appliance or installation pipework exclusively used in a part of premises occupied for non-residential purposes.

"relevant premises" means premises or any part of premises occupied, whether exclusively or not, for residential purposes (such occupation being in consideration of money or money's worth) under –

- *(a) a lease; or*
- *(b) a licence;*

"statutory tenancy" means –

- *(a) in England and Wales, a statutory tenancy within the meaning of the Rent Act 1977 and the Rent (Agriculture) Act 1976; and*
- *(b) in Scotland, a statutory tenancy within the meaning of the Rent (Scotland) Act 1984, a statutory assured tenancy within the meaning of the Housing (Scotland) Act 1988 or a secure tenancy within the meaning of the Housing (Scotland) Act 1987;*

"tenant" means a person who occupies relevant premises being –

- *(a) in England and Wales –*
  - *(i) where the relevant premises are so occupied under a lease, the person for the time being entitled to the term of that lease; and*

Regulation 36(1)–(2)

(ii) where the relevant premises are so occupied under a licence, the licensee;

(b) in Scotland, the person for the time being entitled to the tenant's interest under a lease.

(2) Every landlord shall ensure that there is maintained in a safe condition –

(a) any relevant gas fitting; and
(b) any flue which serves any relevant gas fitting,

so as to prevent the risk of injury to any person in lawful occupation of relevant premises.

ACOP 36(1)–(2)

300 Effective maintenance of appliances usually involves an ongoing programme of regular/periodic inspections, together with any necessary remedial work. In the absence of specific manufacturers' instructions, effective maintenance should include as a minimum:

(a) examination of the physical condition and safe functioning of appliance(s), installation pipework, ventilation and any flue for deterioration;
(b) carrying out performance tests;
(c) taking the necessary remedial action.

Guidance 36(1)–(2)

301 Reference should be made to Appendix 3 and appropriate standards. A list of the relevant standards can be found in the *Legislative, normative and informative document list*.

Regulation 36(3)–(4)

(3) Without prejudice to the generality of paragraph (2) above, a landlord shall –

(a) ensure that each appliance and flue to which that duty extends is checked for safety within 12 months of being installed and at intervals of not more than 12 months since it was last checked for safety (whether such check was made pursuant to these Regulations or not and see regulation 36A);

(b) in the case of a lease commencing after the coming into force of these Regulations, ensure that each appliance and flue to which the duty extends has been checked for safety within a period of 12 months before the lease commences or has been or is so checked within 12 months after the appliance or flue has been installed, whichever is later (and see regulation 36A); and

(c) ensure that a record in respect of any appliance or flue so checked is made and retained until there have been two further checks of the appliance or flue under this paragraph or, in respect of an appliance or flue that is removed from the premises, for a period of 2 years from the date of the last check of that appliance or flue, which record shall include the following information – [1]

(i) the date on which the appliance or flue was checked;
(ii) the address of the premises at which the appliance or flue is installed;
(iii) the name and address of the landlord of the premises (or, where appropriate, his agent) at which the appliance or flue is installed;
(iv) a description of and the location of each appliance or flue checked;
(v) any safety defect identified;
(vi) any remedial action taken;

1 Regulation 36(3) was amended by the Gas Safety (Installation and Use) (Amendment) Regulations 2018.

Regulation 36(3)–(4)

(vii) *confirmation that the check undertaken complies with the requirements of paragraph (9) below;*

(viii) *the name and signature of the individual carrying out the check; and*

(ix) *the registration number with which that individual, or his employer, is registered with a body approved by the Executive for the purposes of regulation 3(3) of these Regulations.*

(4) Every landlord shall ensure that any work in relation to a relevant gas fitting or any check of a gas appliance or flue carried out pursuant to paragraphs (2) or (3) above is carried out by, or by an employee of, a member of a class of persons approved for the time being by the Health and Safety Executive for the purposes of regulation 3(3) of these Regulations.

ACOP 36(3)–(4)

302 A safety check record should not be issued unless and until all the required tests/examinations have been completed on all relevant appliances and flues. A landlord should take prompt action to correct any safety defect shown on a safety check record, which is not remedied at the time of the safety check.

303 Where there is any suspicion that an appliance may be dangerous, the landlord or other responsible person for the premises should ensure that the appliance concerned is not used until the defect(s) have been remedied (see regulation 34).

Regulation 36(5)–(12)

(5) The record referred to in paragraph 3(c) above, or a copy thereof, shall be made available upon request and upon reasonable notice for the inspection of any person in lawful occupation of relevant premises who may be affected by the use or operation of any appliance to which the record relates.

(6) Notwithstanding paragraph (5) above, every landlord shall ensure that –

(a) *a copy of the record made pursuant to the requirements of paragraph 3(c) above is given to each existing tenant of premises to which the record relates within 28 days of the date of the check; and*

(b) *a copy of the last record made in respect of each appliance or flue is given to any new tenant of premises to which the record relates before that tenant occupies those premises save that, in respect of a tenant whose right to occupy those premises is for a period not exceeding 28 days, a copy of the record may instead be prominently displayed within those premises.*

(7) Where there is no relevant gas appliance in any room occupied or to be occupied by the tenant in relevant premises, the landlord may, instead of ensuring that a copy of the record referred to in paragraph (6) above is given to the tenant, ensure that there is displayed in a prominent position in the premises (from such time as a copy would have been required to have been given to the tenant under that paragraph), a copy of the record with a statement endorsed on it that the tenant is entitled to have his own copy of the record on request to the landlord at an address specified in the statement; and on any such request being made, the landlord shall give to the tenant a copy of the record as soon as is reasonably practicable.

Regulation 36(5)–(12)

(8) A copy of the record given to a tenant pursuant to paragraph (6)(b) above need not contain a copy of the signature of the individual carrying out the check if the copy of the record contains a statement that another copy containing a copy of such signature is available for inspection by the tenant on request to the landlord at the address specified in the statement, and on any such request being made the landlord shall make such a copy available for inspection as soon as is practicable.

(9) A safety check carried out pursuant to paragraph (3) above shall include, but shall not be limited to, an examination of the matters referred to in sub-paragraphs (a) to (d) of regulation 26(9) of these Regulations.

(10) Nothing done or agreed to be done by a tenant of relevant premises or by any other person in lawful occupation of them in relation to the maintenance or checking of a relevant gas fitting or flue in the premises (other than one in part of premises occupied for non-residential purposes) shall be taken into account in determining whether a landlord has discharged his obligations under this regulation (except in so far as it relates to access to that gas fitting or flue for the purposes of such maintenance or checking).

(11) Every landlord shall ensure that in any room occupied or to be occupied as sleeping accommodation by a tenant in relevant premises there is not fitted a relevant gas fitting of a type the installation of which would contravene regulation 30(2) or (3) of these Regulations.

(12) Paragraph (11) above shall not apply in relation to a room which since before the coming into force of these Regulations has been occupied or intended to be occupied as sleeping accommodation.

Regulation 36A Determination of date when next safety check is due under regulation 36(3)

Summary of regulation 36A

This regulation sets out when the next safety check must be completed in order to retain the existing deadline date. It also sets out a one-off flexibility that landlords can use to align the date of an appliance check with that of other appliances at the same premises.

Regulation 36A

(1) Where a safety check of an appliance or a flue made in accordance with regulation 36(3)(a) or (b) is or was completed within the period of 2 months ending with the deadline date, that check is to be treated for the purposes of regulation 36(3)(a) and (b) as having been made on the deadline date.

(2) Subject to paragraph (3), the landlord may ensure that an appliance or flue is checked for safety within the 2 month period beginning with the deadline date, instead of checking it within the 12 month period ending with that date.

(3) The discretion conferred by paragraph (2) may be exercised –

- *(a) only once in relation to each appliance or flue in the relevant premises; and*
- *(b) only in order to align the deadline date in relation to the next safety check of that appliance or flue with the deadline date in relation to the next safety check of any other appliance or flue in the same relevant premises.*

Regulation 36A

(4) In this regulation "the deadline date", in relation to a safety check for an appliance or flue, means the last day of the 12 month period within which the check is or was required to be made under regulation 36(3) (a) or (b).[1]

Guidance 36A

304 The changes set out in regulation 36A aim to offer more flexibility in the gas safety checking regime – however, it is not compulsory for landlords to take advantage of this change. If they prefer, landlords can continue with their current regime of gas safety checking, as long as it meets the legal minimum requirements as set out in regulation 36.

305 With the introduction of the new regulation 36A from 6 April 2018 landlords will be able to have gas safety checks carried out any time from 10 to 12 ***calendar months*** after the previous check but still retain the original deadline date as if the check had been carried out exactly 12 months after the previous check.

Record keeping

306 To benefit from this new flexibility and retain the deadline date, the landlord will have to demonstrate that they have complied with the law and carried out the gas safety checks within the required timescales. The record will need to be kept until two further gas safety checks have been carried out.

307 Where a landlord cannot provide the necessary audit trail/documentation, *including the two previous gas safety records*, the expiry date of the current gas safety check will be taken as 12 months from the date of the last gas safety check.

308 This demonstration that they have complied with the law may take the form of a computerised database or a paper file or other means as long as there are records showing the dates of previous gas safety checks, the date of the latest check, and the preserved deadline date (resetting this as and when necessary – see section below on 'resetting the clock'), along with copies of the landlord's gas safety records from the previous two years.

309 There is no legal requirement to include:

(a) an expiry date of the landlord's gas safety record; or
(b) the earliest date you can have your next gas safety check (and retain the deadline date)

on the gas safety records, but landlords may find it helpful to include these details to:

(a) ensure they can demonstrate the necessary audit trail to show that consecutive gas safety checks have been carried out in the prescribed 10–12-month window, thereby retaining the deadline date;
(b) give tenants confidence and clarity over the period of validity of the gas safety record.

310 Even if the engineers include this information on the gas safety records, the legal duty remains with the landlord to be able to demonstrate that gas safety checks have been made within the required timescales.

1 Regulation 36A was added by the Gas Safety (Installation and Use) (Amendment) Regulations 2018

Guidance 36A

Resetting the clock

Gas safety check carried out less than 10 months since the previous check

311 Where a gas safety check is carried out at *less than 10 months* after the previous gas safety check, this will have the effect of 'resetting the clock' and the deadline date will now be 12 months from the date of this latest check.

Gas safety check carried out after the preserved deadline date

312 Where the property remains tenanted it is an offence to have no current gas safety check record in place.

Gas safety check carried out after the preserved deadline date (ie where the property was vacant when the previous gas safety check expired)

313 Where a gas safety check is carried out *more than 12 months* after the previous gas safety check, this will have the effect of 'resetting the clock' and the new deadline date will be 12 months from the date of this later gas safety check.

One-off flexibility to bring dates of safety checks on different gas appliances in the same premises back into line

314 There is a one-off flexibility to allow landlords to realign the date of the annual gas safety check on an appliance with other appliances in a property. It can only be applied once to each appliance, and the maximum amount of time that the gas safety check for the appliance can be extended is 2 months. For example, if an appliance is found to be faulty and is replaced at the time of the actual check, the annual gas safety check on the new appliance can take place for the first time at the same time as the next annual gas safety check on any other appliances in the same property.

PART G Miscellaneous

Regulation 37 Escape of gas

Summary of regulation 37

This regulation details action to be taken by gas suppliers and people responsible for premises in the event of an escape of gas other than natural gas (as covered by the Gas Safety (Management) Regulations 1996). This extends to the emission of, or suspected emission of, carbon monoxide from an appliance using gas, other than natural gas supplied from a network.

Guidance 37

315 This regulation is primarily restricted to LPG and certain other fuel gases (see definition of 'gas' in regulation 2(1)) which do not consist wholly or mainly of methane. It does not cover an escape of natural gas from any network – regulation 37(7) refers. The requirements also extend to an actual or suspected escape or emission of fumes, ie products of combustion containing carbon monoxide (CO), from an appliance using a gas other than natural gas (see regulation 37(8)).

316 Where a gas escape or emission of fumes arises from a dangerous appliance, requirements under regulation 34 also need to be met. Action in the case of a natural gas escape, or suspected emission of CO from a natural gas appliance (where, in either case, natural gas is supplied from a network) is covered under regulation 7 of the Gas Safety (Management) Regulations 1996. Any gas escape or incident involving emission of CO from a dangerous appliance should be reported, as applicable, under the Reporting of Injuries, Diseases and Dangerous Occurrences Regulations 2013 (see summary of legislation at www.hse.gov.uk/gas/domestic/).

317 Regulation 37 extends to certain premises, eg factories, mines and quarries, otherwise excluded from GSIUR (see regulation 2(4)). For the meaning of 'supplier', refer to regulation 2(1)–(2); see also regulation 37(6) concerning sub-contracting to other emergency service providers.

Regulation 37(1)

(1) Where any gas escapes from any pipe of a gas supplier or from any pipe, other gas fitting or gas storage vessel used by a person supplied with gas by a gas supplier, the supplier of the gas shall, within 12 hours of being so informed of the escape, prevent the gas escaping (whether by cutting off the supply of gas to any premises or otherwise).

Guidance 37(1)

318 Gas suppliers/emergency service providers for gas supplies other than natural gas in a network need to make arrangements to ensure that reports of emergencies can be received and responded to 24 hours a day. To avoid delays, responders to emergency calls need an outline script enabling them to:

(a) establish the precise location of the emergency;
(b) in the case of a fuel gas escape:
  (i) establish if and how the leak may be controlled ie by isolating an emergency control or cylinder valve(s), as appropriate;

Guidance 37(1)

- (ii) advise callers how to turn off the gas at the emergency control/cylinder valve(s) and confirm that this has been done, except where this may be dangerous, eg requiring entry into a confined space where there is a smell of gas, when the advice should be not to take such action but instead to vacate the premises;
- (iii) advise callers to open doors and windows to ventilate the property and not to smoke or use electrical appliances/other sources of ignition;

- (c) establish whether there are fumes (escape of CO into the room), and if it is possible, identify the appliance emitting the fumes;
- (d) advise callers, where an escape of CO is suspected, of the immediate steps to be taken, ie to turn off appliances which may be emitting CO (all the appliances in the room/area, where the defective appliance cannot be identified), ventilate the property to disperse CO, and not to use the appliances concerned again until they have been checked by a competent gas engineer who meets the requirements of regulation 3, eg who is Gas Safe registered.

319 In the case of a fuel gas escape, the gas supplier/emergency service provider should attend the emergency as soon as is reasonably practicable after receiving a report of the escape. Once at the emergency, appropriate steps should be taken to bring the situation under control, and make it safe as quickly as possible. The 12-hour period specified in regulation 37(1) is the maximum time that should normally be taken to stop gas escaping.

320 Gas suppliers/emergency service providers need to ensure that they:

- (a) are registered, as appropriate, with a body approved by HSE under regulation 3, eg Gas Safe Register, for work on gas fittings and installation pipework;
- (b) employ an adequate number of competent engineers with sufficient knowledge, appropriate equipment, practical skill and experience to deal with all foreseeable emergency situations (see regulation 3);
- (c) establish written procedures for engineers to follow.

321 The primary duty on gas suppliers/emergency service providers in the event of an emergency is to take action to ensure the situation is made safe. They need to:

- (a) respond to reports of fuel gas escapes and take action as necessary to make the situation safe by preventing gas from escaping; and/or
- (b) respond to reports of suspected or actual CO escapes and advise on the action to be taken to make the situation safe.

322 Calls relating to escapes of gas other than natural gas, eg LPG, may be misdirected to the National Gas Emergency Service. Gas suppliers may wish to consider arrangements for liaison with National Grid plc in respect of such calls.

Regulation 37(2)–(4)

(2) If the responsible person for any premises knows or has reason to suspect that gas is escaping into those premises, he shall immediately take all reasonable steps to cause the supply of gas to be shut off at such place as may be necessary to prevent further escape of gas.

(3) If gas continues to escape into those premises after the supply of gas has been shut off or when a smell of gas persists, the responsible person for the premises discovering such escape or smell shall immediately give notice of the escape or smell to the supplier of the gas.

Regulation 37(2)–(4)

(4) Where an escape of gas has been stopped by shutting off the supply, no person shall cause or permit the supply to be re-opened (other than in the course of repair) until all necessary steps have been taken to prevent a recurrence of such escape.

Guidance 37(2)–(4)

323 Where a fuel gas escape is suspected or known to be occurring, the responsible person (see definition in regulation 2(1)) should take immediate action to cause the emergency control or gas cylinder shut-off valve(s), as appropriate, to be closed, except where this would be dangerous, eg requiring entry into a confined space where there is a smell of gas when the premises should instead be evacuated. If gas continues to escape, the responsible person should immediately notify the supplier emergency gas service that there is a suspected escape of gas.

324 In any case of a gas escape, doors and windows should be opened to ventilate the property and all sources of ignition, eg smoking or use of electrical equipment, should be avoided. The gas supply should not be reinstated until remedial action has been taken and appropriate checks made (eg that appliance isolation valves are closed, as appropriate) to avoid a further escape of gas.

325 In the case of a suspected escape of CO the responsible person should turn off any appliances suspected of emitting CO, open doors and windows to ventilate the property and contact the gas supplier emergency service immediately. However, where a particular appliance is known to be the source of the escape, the responsible person should ensure that it is not used and contact someone who is, or is employed by, a member of an HSE-approved class of persons under regulation 3(3) (eg a Gas Safe registered engineer) to repair, replace or otherwise make safe the appliance (see also regulation 34).

326 The name and means of contacting the supplier emergency gas service may be found on a notice near the meter, or where there is no meter, the emergency gas control (see regulations 9 and 15, and associated guidance). For LPG, supplier information may be found on the bulk tank or cylinder. If there is no notice, or the relevant information is not given, details of the gas supplier may be found in the telephone directory or in the useful contact numbers section of www.gassaferegister.co.uk.

327 Similar action to the above is required under the Gas Safety (Management) Regulations 1996 for any natural gas escape or suspected emission of CO from a natural gas appliance (where the natural gas is supplied from a network). In such cases, the National Gas Emergency Service Freephone Number 0800 111 999 should be contacted.

Regulation 37(5)

(5) In any proceedings for an offence under paragraph (1) above it shall be a defence for the supplier of the gas to prove that it was not reasonably practicable for him effectually to prevent the gas from escaping within the period of 12 hours referred to in that paragraph, and that he did effectually prevent the escape of gas as soon as it was reasonably practicable for him to do so.

Guidance 37(5)

328 There may be circumstances, eg a major leak from a storage tank, when it may not be feasible to stop the gas escape within 12 hours. To meet the requirements of this regulation, gas suppliers will need to ensure that the situation is brought under control and made safe as soon as is reasonably practicable. Where it is not possible to stop leakage within 12 hours, gas suppliers/emergency service providers will need to demonstrate that they took all reasonable steps to do so.

Regulation 37(6)

(6) Nothing in paragraphs (1) and (5) above shall prevent the supplier of the gas appointing another person to act on his behalf to prevent an escape of gas supplied by that supplier.

Guidance 37(6)

329 Gas suppliers may subcontract emergency service provision to another person or organisation (the 'emergency service provider'), eg in the interests of providing a more local and immediate response in a particular area or at a specific site. This may, for instance, be especially relevant for local companies supplying gas in cylinders which may be used over a wide geographical area. However, where an emergency service provider is appointed in this way, the primary responsibility remains with the supplier concerned (see paragraphs 47–51 for allocation of duties in particular circumstances).

Regulation 37(7)

(7) Nothing in paragraphs (1) to (6) above shall apply to an escape of gas from a network (within the meaning of regulation 2 of the Gas Safety (Management) Regulations 1996) or from a gas fitting supplied with gas from a network.

Guidance 37(7)

330 This regulation limits requirements in regulation 37(1)–(6) primarily to LPG and certain other gases (see definition of 'gas' in regulation 2(1)) which are not wholly or mainly composed of methane. The exclusion of an escape of natural gas from any 'network' effectively means that virtually all escapes of natural gas will not be covered under this regulation; the only exception is where such an escape occurs from a 'stand-alone' network of pipes which conveys gas only to non-domestic premises falling within GSIUR, eg an office block.

331 In all other cases, escapes of natural gas are covered by requirements under regulation 7 of the Gas Safety (Management) Regulations 1996. Action in the case of an emission or suspected emission of CO from a natural gas appliance is similarly covered.

Regulation 37(8)

(8) In this regulation any reference to an escape of gas from a gas fitting includes a reference to an escape or emission of carbon monoxide gas resulting from incomplete combustion of gas in a gas fitting, but, to the extent that this regulation relates to such an escape or emission of carbon monoxide gas, the requirements imposed upon a supplier by paragraph (1) above shall, where the escape or emission is notified to the supplier by the person to whom gas has been supplied, be limited to advising that person of the immediate action to be taken to prevent such escape or emission and the need for the examination and, where necessary, repair of the fitting by a competent person.

Guidance 37(8)

332 The requirements of regulation 37 apply both to an escape of fuel gas (other than any escape of natural gas from a 'network', as covered by the Gas Safety (Management) Regulations 1996), and to suspected or actual emissions of fumes, ie CO, from an appliance using gas (other than natural gas supplied from a 'network').

333 This means that the requirements in regulation 37 extend to an emission or suspected emission of CO from any appliance using 'gas' as defined in regulation 2(1) (eg LPG) apart from (in most cases) those using natural gas – see regulation 37(7).

334 However, the duty on gas suppliers in regulation 37(1) with regard to a report of CO escape/emission is restricted to providing advice on how to prevent the escape/emission (including identification of the faulty appliance/fitting (if possible) and action to isolate the gas supply) and on the need for examination and (where necessary) repair of the appliance/fitting by a competent person, eg a Gas Safe registered engineer (see regulation 3).

Regulation 38 Use of antifluctuators and valves

Summary of regulation 38

This regulation outlines protective measures, as stipulated by the gas transporter, to be taken by a consumer where gas is used with plant (such as a compressor or engine) liable to cause dangerous fluctuation of pressure in the gas supply, or where an extraneous gas (eg compressed air) is used in connection with the consumption of gas.

Regulation 38(1)–(4)

(1) Where a consumer uses gas for the purpose of working or supplying plant which is liable to produce pressure fluctuation in the gas supply such as to cause any danger to other consumers, he shall comply with such directions as may be given to him by the transporter of the gas to prevent such danger.

(2) Where a consumer intends to use for or in connection with the consumption of gas any gaseous substance he shall –

(a) give to the transporter of the gas at least 14 days notice in writing of that intention; and
(b) during such use comply with such directions as the transporter may have given to him to prevent the admission of such substance into the gas supply;

and in this paragraph "gaseous substance" includes compressed air but does not include any gaseous substance supplied by the transporter.

(3) Where a direction under paragraphs (1) or (2) above requires the provision of any device, the consumer shall ensure that the device is adequately maintained.

(4) Any direction given pursuant to this regulation shall be in writing.

Guidance 38(1)–(4)

335 This regulation applies to consumers who use plant that may create pressure fluctuation in the gas supply (eg a compressor, engine or similar apparatus liable to produce a pressure below atmospheric in the gas main), or who intend to use in conjunction with gas another gaseous substance such as compressed air (for instance in gas/air mixing equipment or other process plant). The consumer should notify the gas transporter of intentions in either case (at least 14 days' notice is specifically required under regulation 38(2)(a)), and is required to comply with directions given by the transporter to protect the gas supply against possible pressure fluctuation, or introduction of extraneous gas, to prevent danger to other consumers. Any device, such as an antifluctuator or non-return valve, specified by the gas transporter, needs to be provided and maintained by the consumer.

336 Under schedule 28 of the Gas Act 1986 (as inserted by the Gas Act 1995 and amended by these regulations), the gas transporter may disconnect premises where a consumer fails to comply with the requirements in regulation 38. The gas transporter is also empowered to disconnect, remove, test and replace any appliance/device required under this regulation, and to claim expenses from the consumer where the appliance is not found to be in proper order and repair.

Regulation 39 Exception as to liability

Regulation 39

No person shall be guilty of an offence by reason of contravention of regulation 3(2) or (6), 5(1), 7(3), 15, 16(2) or (3), 17(1), 27(5), 30 (insofar as it relates to the installation of a gas fire, other gas space heater or a gas water heater of more than 14 kilowatt gross heat input), 33(1), 35 or 36 of these Regulations in any case in which he can show that he took all reasonable steps to prevent that contravention.

Guidance 39

337 This regulation provides certain exceptions as to liability under which a person is not deemed to be guilty of an offence where they can show that they took all reasonable steps to prevent contravention of the provision concerned. This publication contains guidance in various areas, eg concerning access to premises to discharge landlords' duties under regulation 36. However, it is for a court to decide whether 'all reasonable steps' have in fact been taken in particular circumstances, and whether a person is guilty of an offence.

338 The exceptions as to liability only apply to the particular provisions specified in regulation 39. They do not extend to the prohibition in regulation 30 concerning installation of certain instantaneous water heaters.

Regulation 40 Exemption certificates

Regulation 40

(1) Subject to paragraph (2), the Health and Safety Executive may, by a certificate in writing, exempt any person or class of persons from any requirement or prohibition imposed by these Regulations, and any such exemption may be granted subject to conditions and to a limit of time and may be revoked at any time by a certificate in writing.

(2) The Health and Safety Executive shall not grant any such exemption unless, having regard to the circumstances of the case and in particular to –

- *(a) the conditions, if any, which it proposes to attach to the exemption; and*
- *(b) any other requirements imposed by or under any enactment which apply to the case,*

it is satisfied that the health and safety of persons likely to be affected by the exemption, will not be prejudiced in consequence of it.

Guidance 40

339 This regulation enables HSE to grant exemptions. The intention is to use this power only in exceptional circumstances, such as to deal with circumstances which could not be foreseen when GSIUR was made (eg arising from technological change/innovation).

340 Regulation 40(2) requires HSE to ensure that standards of health and safety will not be prejudiced by any exemption, so it is not sufficient for an application for exemption simply to assert that it is inconvenient or costly to comply. The applicant needs to explain carefully which particular provision cannot be observed and why, and the means by which an equivalent standard of health and safety is to be maintained.

Regulation 40A Review

Regulation 40A

(1) The Secretary of State must from time to time –

(a) carry out a review of the regulatory provision contained in regulations 2(4)(g), 26(9)(c) and (ca), 36(3) and 36A; and
(b) publish a report setting out the conclusions of the review.

(2) The first report must be published before 6th April 2023.

(3) Subsequent reports must be published at intervals not exceeding 5 years.

(4) Section 30(4) of the Small Business, Enterprise and Employment Act 2015 requires that a report published under this regulation must, in particular –

(a) set out the objectives intended to be achieved by the regulatory provision referred to in paragraph (1)(a);
(b) assess the extent to which those objectives are achieved;
(c) assess whether those objectives remain appropriate; and
(d) if those objectives remain appropriate, assess the extent to which they could be achieved in another way which involves less onerous regulatory provision.

(5) In this regulation, "regulatory provision" has the same meaning as in sections 28 to 32 of the Small Business, Enterprise and Employment Act 2015 (section 32 of that Act).[1]

Regulation 41 Revocation and amendments

Regulation 41

(1) The Gas Safety (Installation and Use) Regulations 1994, the Gas Safety (Installation and Use) (Amendment) Regulations 1996 and the Gas Safety (Installation and Use) (Amendment) (No. 2) Regulations 1996 are hereby revoked.

(2) Schedule 2B to the Gas Act 1986 shall be amended as follows –

(a) In paragraph 17(1) the words "pressure fluctuation in the transporter's pipe-line system and any other" and the words "or danger" shall be deleted;
(b) In paragraph 17(2) after the words "if so required" there shall be added "other than for the purpose of preventing danger"; and
(c) In paragraph 17(5) and (6) after the words "this paragraph" there shall be added "or regulation 38 of the Gas Safety (Installation and Use) Regulations 1998 or directions made thereunder".

1 Regulation 40A was added by the Gas Safety (Installation and Use) (Amendment) Regulations 2018.

Appendix 1 Notice of approval

By virtue of section 16(4) of the Health and Safety at Work etc Act 1974, and with the consent of the Secretary of State for Work and Pensions, the Health and Safety Executive has on 25 September 2013 approved the revised Code of Practice entitled *Safety in the installation and use of gas systems and appliances* (fourth edition, 2013, L56).

The revised Code of Practice gives practical guidance on the Gas Safety (Installation and Use) Regulations 1998 and, in Part B, regulation 3, on section 2(1) and (2)(c) and section 3(1) and (2) of the 1974 Act with regard to standards of training in safe gas installation.

By virtue of section 16(5) and with the consent of the Secretary of State for Work and Pensions under that paragraph, the Health and Safety Executive has withdrawn its approval of the Code of Practice entitled *Safety in the installation and use of gas systems and appliances*, which came into effect on 31 October 1998 which shall cease to have effect on 11 November 2013.

The Code of Practice comes into effect on 11 November 2013.

Sue Johns
Secretary to the Board of the Health and Safety Executive
7 November 2013

Appendix 2 Diagrams of typical installations

Guidance

1 These simplified installation diagrams are only intended for use as an aid to identify and indicate the relative positions of components described in the regulations, ACOP and guidance.

2 The diagrams are only general illustrations of typical arrangements and variations may occur in specific circumstances. For instance, in certain cases (eg Figures 8 and 10) the 'remote' valve shown as the first emergency control downstream of the distribution main will not be an emergency control in cases where it is intended for use by the gas transporter and/or emergency services rather than gas consumers. In such cases, a valve or valves downstream (eg within individual dwellings) will act as the first emergency control and define the boundary between the service pipe and installation pipework. Reference should also be made to the Institution of Gas Engineers and Managers' (IGEM) document IGEM/G/1 *Defining the end of the network, a meter installation and installation pipework*, which provides detail on a wide range of standard and legacy gas supply installation arrangements.

3 The upstream boundary of the service pipe (ie with the distribution main) in Figures 5–11 is not shown.

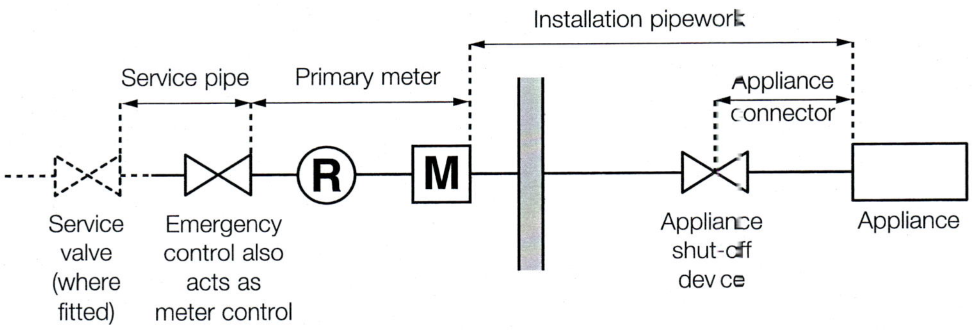

Figure 5 Outside meter installation

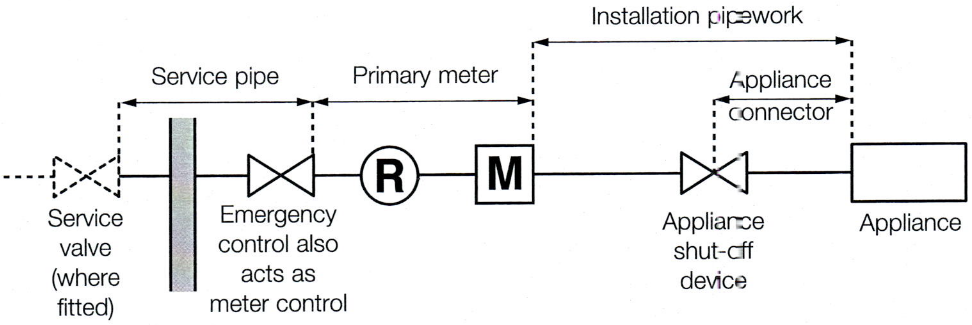

Figure 6 Inside meter installation

Guidance

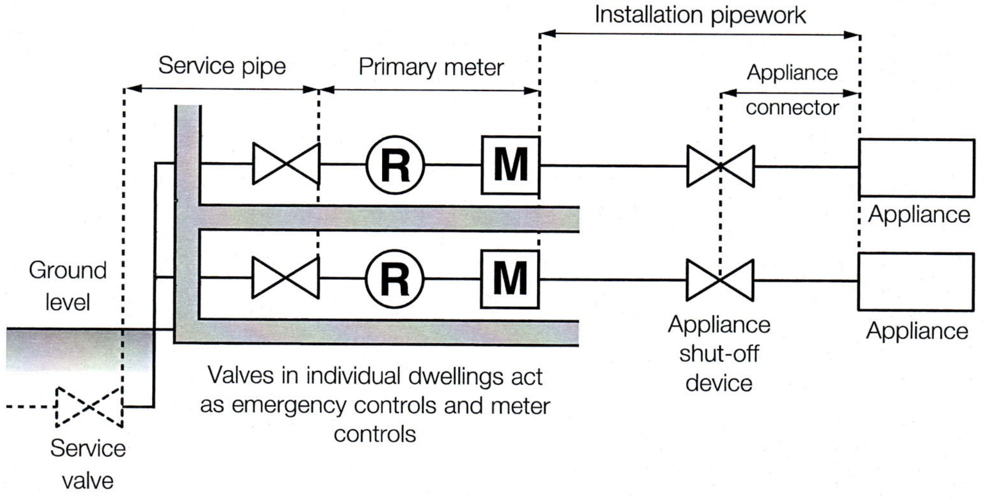

Figure 7(a) Multi-occupancy installation – external riser

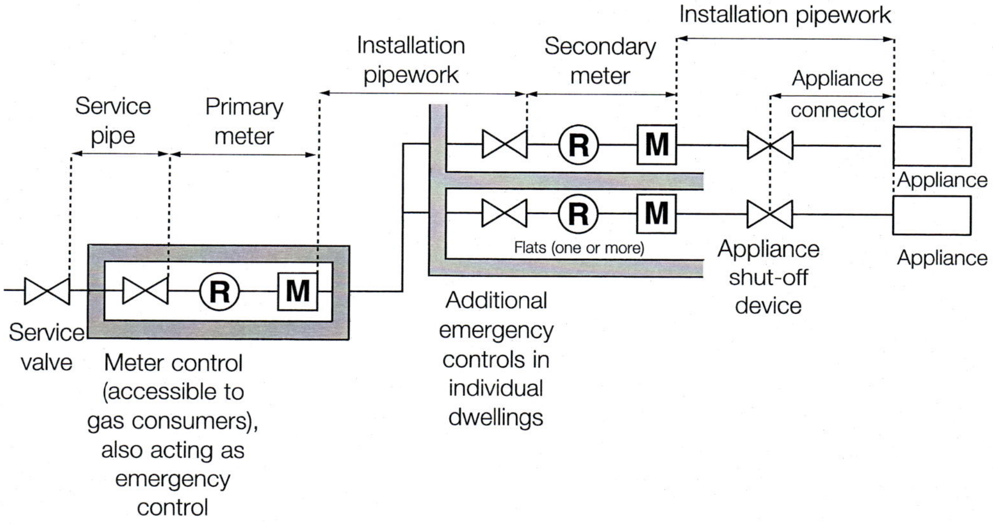

Figure 7(b) Multi-occupancy installation (remote bulk meter)

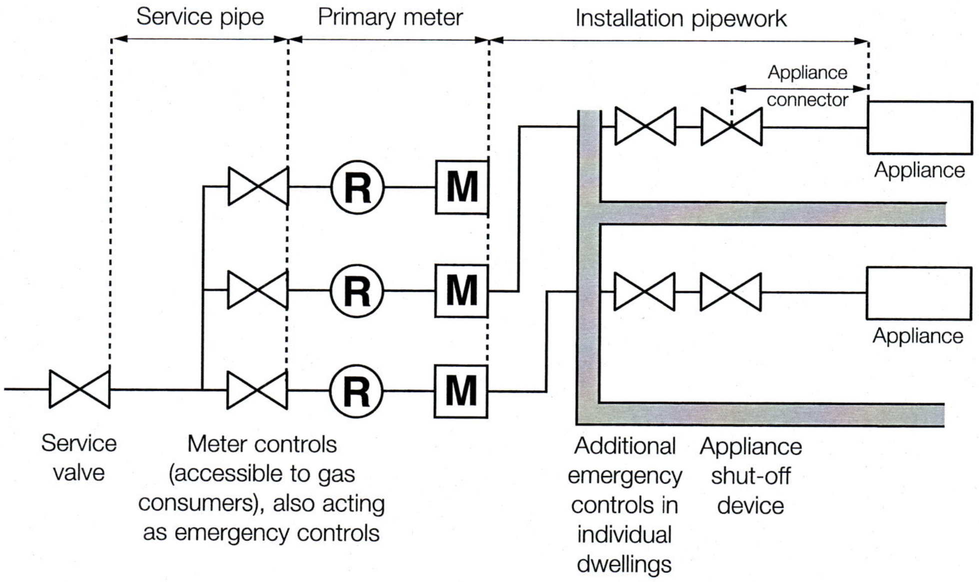

Figure 8 Multi-occupancy installation (remote meters)

Guidance

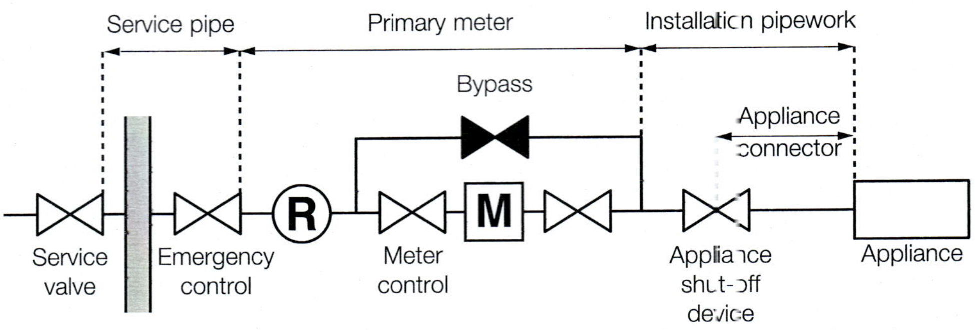

Figure 9 Meter with bypass, eg industrial/commercial

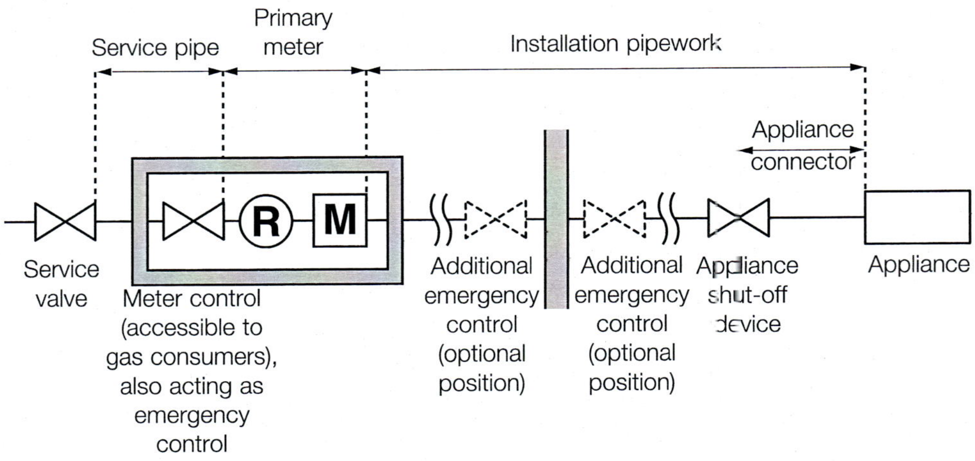

Figure 10 Meter remote from premises, eg meter houses

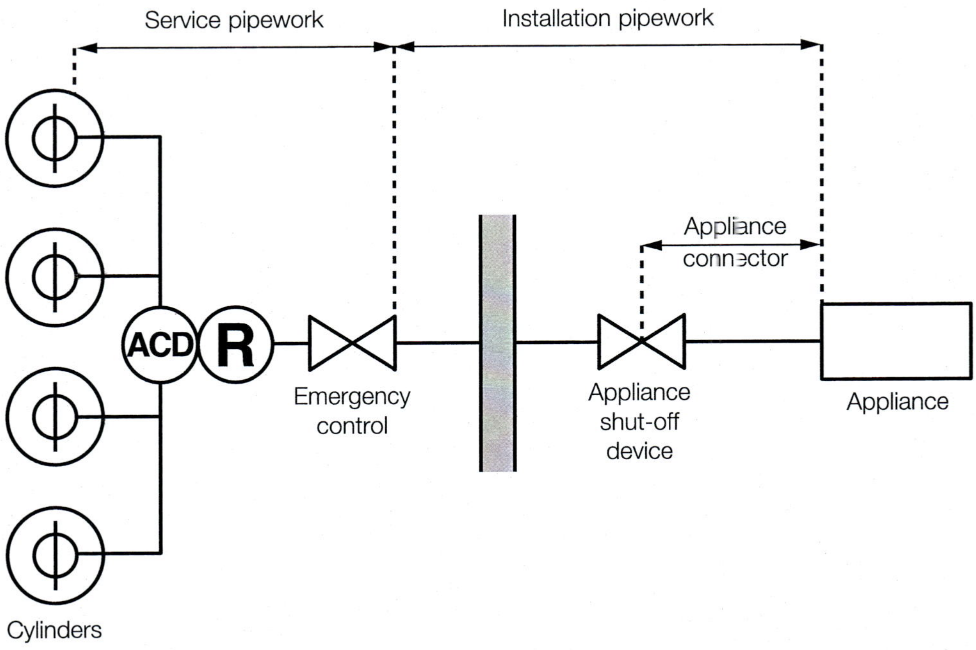

Figure 11(a) LPG systems – cylinder installation

Guidance

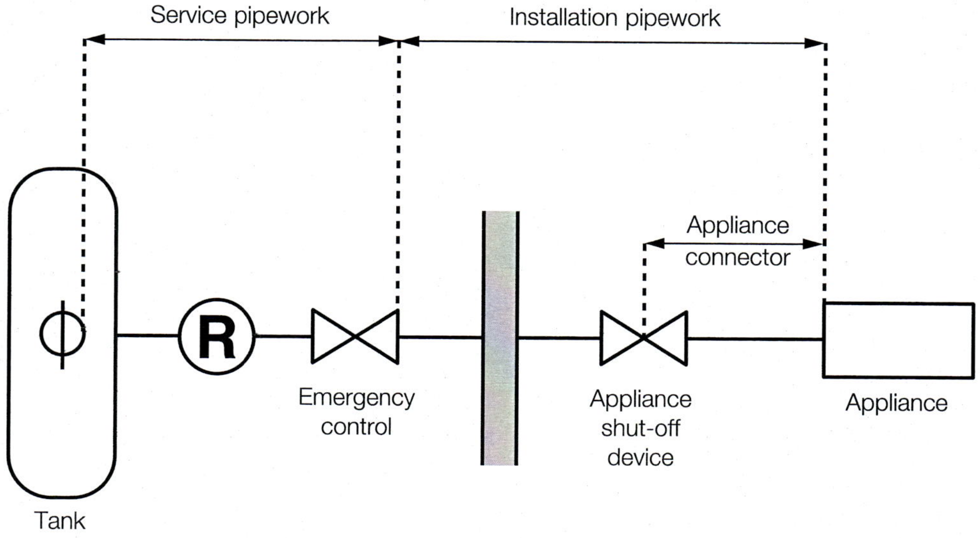

Figure 11(b) LPG systems – single bulk tank installation
(Note: Some vapour installations may have an additional regulator)

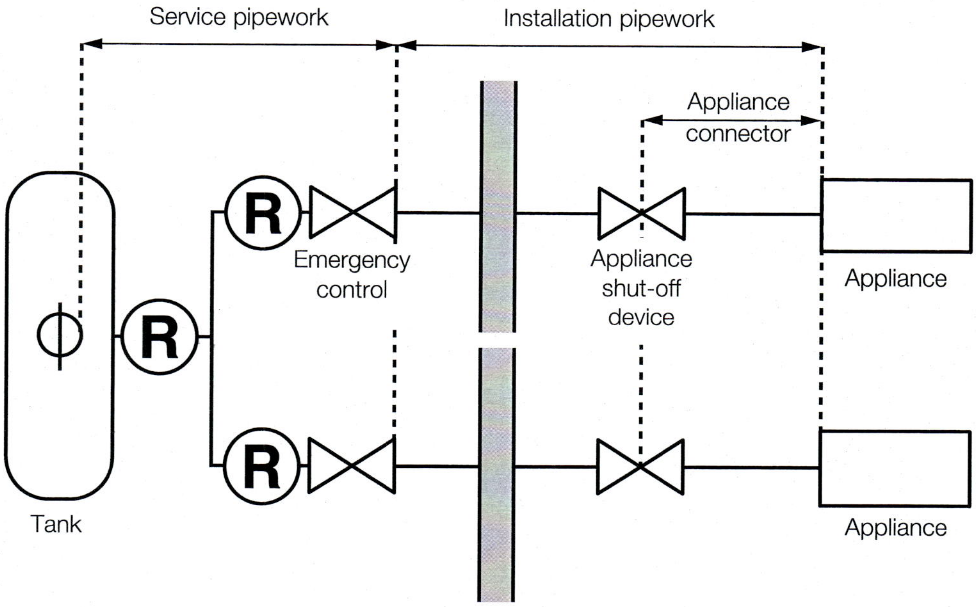

Figure 11(c) LPG systems – single bulk tank installation (multi-residence)

R — Regulator

M — Meter

Gas emergency control – situated as near as is reasonably practicable to the point where the pipe supplying gas enters the premises or building (see also paragraph 2 of this appendix)

Vapour valve on the storage vessel

ACD — Automatic change-over device

Appendix 3 Requirements for appliances and flues

Guidance

1 This appendix summarises the main points that need to be addressed, as appropriate, in installation, commissioning, maintenance or safety check work on gas appliances and flues/chimneys serving those appliances. Additional matters may require attention in particular circumstances; reference should be made to the relevant legislation (eg Building Regulations), manufacturers' instructions and appropriate standards, where applicable, for further information.

Appliances

2 Gas engineers carrying out installation, commissioning, maintenance or safety check work should ensure that:

- (a) the appliance is suitable and, if necessary, adjusted for the type of gas being supplied;
- (b) the appliance type is suitable for the room/location in which it is installed;
- (c) the room or space where the appliance is located is adequately ventilated, and that the means of ventilation is suitable to provide a sufficient permanent supply of air to the appliance for safe combustion of gas. Account needs to be taken of:
  - (i) the volume of the room or space;
  - (ii) whether draught exclusion or new windows have been fitted, or other changes that could affect ventilation have been made to the room containing the gas appliance(s);
  - (iii) the location, type, size and configuration of airbricks and other permanent air vents;
  - (iv) whether air inlet ducts of room-sealed appliances are correctly located and remain unobstructed; and
  - (v) any other factors which could affect the adequacy of the air supply;
- (d) there are adequate and suitable means for removing products of combustion from the appliance (see requirements in 'Flues/chimneys' in paragraphs 3–4);
- (e) where it can be measured, the operating pressure and/or heat input of the appliance is correct. Information relating to the appropriate appliance settings is given in manufacturers' installation/servicing instructions or on the appliance data badge;
- (f) combustion performance is measured where specified by the manufacturer;
- (g) the flame picture (ie appearance) is visually checked where possible to ensure that it is satisfactory;
- (h) the appliance and associated gas fittings/safety devices (eg flame protection and fan proving systems) function safely;
- (i) any case seals are forming an effective seal;
- (j) the appliance is physically stable, securely fitted and properly connected to other gas fittings;
- (k) the appliance is gas-tight.

Guidance

Flues/chimneys

3 Gas engineers testing and checking flues/chimneys (to determine whether they are suitable and in a proper condition for safe operation of the appliance they are intended to serve) should ensure that the requirements in paragraphs 5–20 below are met.

4 How this is carried out will, to some extent, depend on the particular circumstances, and is a matter for judgement by the competent gas engineer carrying out the work. In some cases, the equipment may need to be dismantled, eg to establish continuity of a flue/chimney path.

General requirements

5 The flue/chimney should be examined to confirm it is complete, continuous and correctly installed throughout its length. It should be undamaged and adequately supported, paying particular attention to any joints and sections of the flue/chimney passing through any roofspace or void.

6 For an existing flue/chimney, where access cannot be gained for examination (eg where it passes through adjacent property), other means of managing the risk of flue system failure should be adopted. Relevant industry guidance should be followed and appropriate action taken.

7 Where there are good reasons to suspect flueing problems (eg because of flue flow/spillage test failure, ceiling staining or any evidence of products of combustion entering the property etc) the complete length of the flue/chimney should be examined. Any associated appliance must not be used unless or until this is done and it has been confirmed safe to use.

Room-sealed flues

8 The terminal should be correctly sited. This should include adequate separation from any site boundary, openable windows and doors, air vents, other terminals etc to ensure that the flue/chimney will continue to function safely and emissions will not present a hazard to any person either on the site concerned or in an adjoining property, including in the event of any future building development on adjoining premises (which might extend close or up to the property boundary).

9 Where access cannot be gained for examination, such as where a flue/chimney runs partly through property not owned by the responsible person/dutyholder (eg landlord) and access is refused by the occupant, all reasonable steps should be taken to ensure overall flue/chimney integrity. This could reasonably involve making enquiries in writing or in person with occupants of other properties and requesting to see evidence of examinations.

Open/closed flues

10 A flue/chimney should not have openings, other than:

- (a) a draught diverter, balancing damper or relief opening which is in the same room, or space, as the appliance being served;
- (b) an opening for inspection or cleaning, which is fitted with a non-combustible gas-tight cover; or
- (c) an opening integral and essential to the correct operation of the flue/chimney, which is purpose-designed and properly located for its application, eg an opening in a vertex flue in an unoccupied roofspace (a guard should be fitted around such an opening, where necessary, to prevent restriction or possible ignition of nearby materials).

Guidance

11 No more than one appliance should be connected to a flue/chimney, unless it has been specifically designed for this purpose.

12 The effect of other flued appliances or fans etc in the same room should be properly taken into account. Where more than one such appliance or fan is installed it should be ensured that they are working safely with all permutations of appliances that may be in operation at any one time, including those burning fuel other than gas.

13 The terminal should be correctly sited; this should include adequate separation from other nearby structures, eg openable windows, chimney stacks etc.

14 Any flue/chimney should not be (or have been) closed over/modified so as to interfere with flue/chimney operation. Any flue damper or restrictor plates should be removed or permanently fixed open to avoid obstructing the flue/chimney.

15 Any debris, eg in a fire/catchment space, should be removed before installing/reinstalling an appliance. The catchment space should be of adequate size and any openings within this space (other than those for the appliance connection and flue/chimney) should be effectively sealed. Flues/chimneys previously used for oil or solid fuel appliances need to be effectively swept and examined before being used with a gas appliance.

16 Where a masonry chimney seems to be in a poor condition or poorly constructed, it should be thoroughly examined to determine whether it is safe for continued use. Remedial work should be carried out where necessary and consideration given to fitting a (correctly sized) flue/chimney liner; in certain cases, a liner is required under Building Regulations.

17 Where components such as external sections of flues/chimneys (eg terminals), liners and inspection openings show signs of damage (eg corrosion), they should be replaced as necessary. When replacing an appliance connected to a chimney lined with a liner, the existing liner should normally be replaced unless it is considered suitable for safe operation throughout the lifespan of the new appliance.

18 When flue/chimney liners are fitted or replaced, the flue/chimney should be tested to ensure it is operating satisfactorily.

19 A flue flow test should be carried out to check the effectiveness of the flue/chimney and ensure that there is no leakage into another part of the premises (including any roofspace or void) or, as appropriate, other adjoining premises. (This is particularly important where a number of chimneys combine into a multiple chimney stack.) Smoke coming out of a terminal which is not the correct terminal, or a down-draught or no-flow condition, indicates an unsatisfactory flue/chimney.

20 After completion of a satisfactory flue flow test, a spillage test should be carried out, with the appliance connected and operating, to check that the products of combustion are being safely removed. The tests should be conducted in accordance with the appliance manufacturer's instructions, and take into account any factors which may adversely affect flue/chimney performance, eg operation of fans, tumble driers or forced-air heating systems in the room housing the appliance or any adjoining room (with interconnecting doors open).

Note: Commercial (and industrial) premises often have purpose-designed flue/chimney systems. The flue/chimney system's manufacturer's instructions and/or appropriate standards need to be followed when maintaining and examining such systems.

Glossary

appropriate standard Any reference to an 'appropriate standard' is a reference to any of the following which are current at the time of the work activity:

(a) a British Standard (see References);
(b) a relevant standard or code of practice of a national standards body of any member state of the European Union;
(c) a relevant technical specification acknowledged for use as a standard by a public authority of any member state of the EU;
(d) traditional procedures of manufacture of a member state of the EU where these are the subject of a written technical description sufficiently detailed to permit assessment of the goods or materials for the use specified;
(e) a specification sufficiently detailed to permit assessment for goods or materials of an innovative nature (or subject to an innovative process of manufacture such that they cannot comply with a recognised standard or specification) and which will fill the purpose provided by the specified standard;

provided that the proposed standard, code of practice, technical specification or procedure of manufacture provides, in use, equivalent levels of safety, suitability and fitness for purpose to those achieved by the standard to which it is expressed to be equivalent.

The mutual recognition in the previous paragraph is also extended to products originating in European Economic Area member states who are contracting parties to the Agreement on the European Economic Area, which conform to the standards, regulations, specifications or traditional procedures of manufacture legally applied in those states and which offer an equivalent standard of safety to that required in GSIUR.

basement Any part of a building that is partly or wholly underground.

Building Regulations Unless the context otherwise requires, reference to 'Building Regulations' means those Regulations currently in force in England and Wales as well those applicable in Scotland (see www.gassaferegister.co.uk for details of relevant Building Regulations).

dutyholder A person who has duties under these Regulations.

gas engineer The terms 'gas engineer' and 'engineer' are not defined in GSIUR but are used in this guidance and ACOP to mean any person who carries out work on a gas system or gas storage vessel. This includes (but is not restricted to) any person who installs, services, maintains, connects, reconnects, disconnects, removes or repairs gas appliances or other fittings whether they are an employer, employee, self-employed or working on their own behalf, ie in a 'do-it-yourself' activity.

gas system The term 'gas system' as used in this ACOP means all gas installation pipework, fittings and appliances as well as the provision of ventilation and flueing.

For the purposes of this publication, where appliance standards refer to domestic use, this should be taken as reference to the type of appliance rather than the location. A small cooker in an office canteen would therefore normally be regarded as a domestic-type appliance in this context.

in writing Where the phrase 'in writing' is used in this document, this can mean both by traditional hard copy or by email.

secondary meter Any reference to a 'secondary meter' includes a 'sub-deduct meter' (see paragraph 196).

References

1 A list of appropriate Standards, ACOPs, guidance and other relevant information sources can be found on the Gas Safe Register website in the *Legislative, normative and informative document list* at www.gassaferegister.co.uk

2 Details of the Canal & River Trust/Environment Agency Boat Safety Scheme is available from the Boat Safety Scheme at www.boatsafetyscheme.org. In the case of inland waterways not covered by this scheme, advice on safety requirements should be sought from the navigation authority concerned.

3 UKAS is the UK's National Accreditation Body. There is more information at www.ukas.com

4 *Standards of training in gas work* Institution of Gas Engineers and Managers (IGEM) www.igem.org.uk

5 *The Gas Industry Unsafe Situations Procedure* www.gassaferegister.co.uk

6 There is information about UKLPG at www.uklpg.org or email mail@uklpg.org

7 Building Regulations www.planningportal.gov.uk

8 *Safe use of work equipment. Provision and Use of Work Equipment Regulations 1998. Approved Code of Practice and guidance* HSE 2014 www.hse.gov.uk/pubns/books/l22.htm

Further information

For information about health and safety visit https://books.hse.gov.uk or http://www.hse.gov.uk.

You can view HSE guidance online and order priced publications from the website. HSE priced publications are also available from bookshops.

To report inconsistencies or inaccuracies in this guidance email: commissioning@williamslea.com.

British Standards can be obtained in PDF or hard copy formats from BSI: http://shop.bsigroup.com or by contacting BSI Customer Services for hard copies only. Tel: 020 8996 9001 email: cservices@bsigroup.com.

The Stationery Office publications are available from The Stationery Office, PO Box 29, Norwich NR3 1GN Tel: 0333 202 5070 Fax: 0333 202 5080 email: customer.services@tso.co.uk Website: www.tsoshop.co.uk/ (They are also available from bookshops.)

Statutory Instruments can be viewed free of charge at www.legislation.gov.uk where you can also search for changes to legislation.

IGEM standards can be obtained in hard copy formats from IGEM www.igem.org.uk or by contacting IGEM Tel: 01889 561431 email: igem@yesresponse.co.uk.

General advice on gas safety issues may be obtained from HSE at www.hse.gov.uk or from the Gas Safe Register on 0800 408 5500 (www.gassaferegister.co.uk)

Information, including on assessment standards, accredited certification bodies and training providers is available from Energy and Utility Skills www.euskills.co.uk

Further guidance on gas (and LPG) safety in catering can be obtained from the National Catering Association NCASS at www.ncass.org.uk